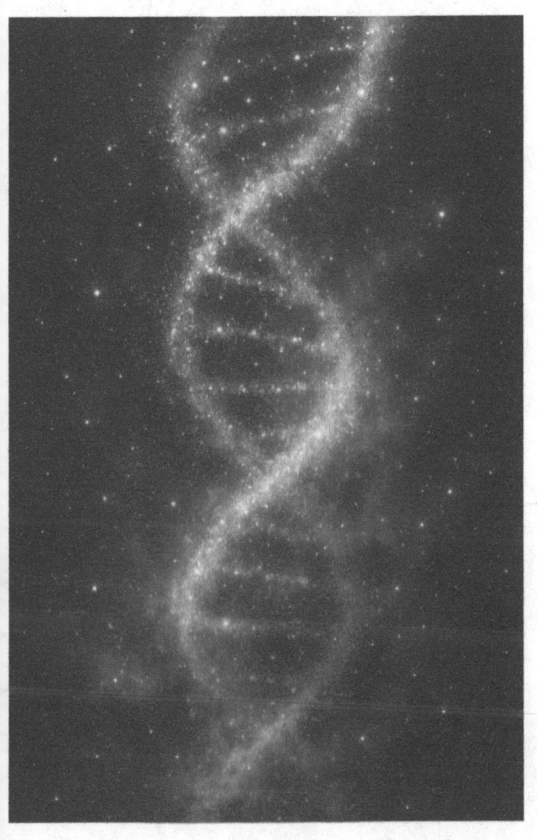

A concise, evidence-based exploration showing how Genesis aligns with modern science. Life and the cosmos aren't accidents—they are intentional. Creation wasn't random. It was written—by design.

Dustin Bullard

BY DESIGN: HOW SCIENCE PROVES THE HAND OF GOD

BOOK 1: CODE AND COSMOS

Printed in the United States of America

ISBN: 978-1-968013-00-4

Cover design by Dustin Bullard Interior design and formatting by Dustin Bullard

Published by 5²40 Project

First Edition

By Design: How Science
Proves the Hand of God
Book 1: Code and Cosmos
1

By Design: How Science Proves the Hand of God
Book 1: Code and Cosmos

Author Introduction

Dustin Bullard

I didn't grow up in a home—I grew up in houses I couldn't call mine. My parents divorced when I was young, and I spent my childhood bouncing between them, but neither place ever felt like I belonged.

In one home, the attention always went to my stepfather's children. In the other, my stepmother's mood swings ruled our lives. I endured punishments no child should face—times when discipline became cruelty. I was beaten for bedwetting. Shamed. Humiliated. I was once struck across the face while wearing a sling for a broken arm—because of an "F" on a report card. There were long car rides that started with smiles and ended in violence. Even my father, though he never hit me, often cut me down with words. I always knew I came second—to someone else's relationship, someone else's peace.

And so I learned to be invisible. I loved science fiction. I climbed trees to hide. I used a toy spy gun to listen through walls—because even when adults tried to hide the yelling, I still wanted to understand. There were glimmers of joy—friends, toys, laughter—but the core memory of my childhood was *displacement*.

And yet, I still hoped life had meaning. That maybe, just maybe, there was *design*.

I met my wife in high school. She was the most beautiful girl I'd ever seen—kind, real, and wounded like me. We bonded through our brokenness. Our family lives had been chaotic, and we found comfort in each other. We married young. We had children young. We moved across states, across countries—Mississippi, Idaho, Texas, Kentucky, even Abu Dhabi.

I worked hard. I sacrificed a lot to provide for my family—even lived in a tent in Kandahar, Afghanistan, washing dishes to pay the bills. But I wasn't present. Not really. I was a provider, but not a partner. We both poured everything into our children— so much so that we left nothing for each other. And we left God out completely.

That changed when my wife gave her life to Christ. She was baptized while I was working in another state. And I didn't rejoice—I got angry. I felt like she'd made a major decision without me. Like she'd let other people speak into her life when I should have been the one leading. I pulled away. I shut down.

And then, one Sunday, I sat beside her at church. I didn't expect anything to happen. But the pastor was preaching from the Sermon on the Mount—and when he read, *"But I say unto you, That whosoever is angry with his brother without a cause shall be in danger of the judgment"* (Matthew 5:22)—it pierced me. That was me. I had carried anger my whole life. Anger at my parents. At my stepparents. At my wife. At God. And that moment, something broke.

3

I gave my life to Jesus. I was baptized. And then… the real battle began.

Shortly after my baptism, I fell into the darkest chapter of my life— infidelity. I justified it because our marriage had been so cold. A voice inside me—*that sounded like me*—kept telling me I deserved it. That if she loved me, she'd show it more. That I had every right. I made secret rules she couldn't meet—rules she didn't even know existed.

But there was another voice too. One that whispered truth. It was quieter. Farther away. It didn't sound like me, so I ignored it.

I stayed in that state for two years. I wore the mask of a Christian. But I was dying inside. I seriously considered ending my life. Not once—*multiple* times. Even though I had claimed Christ, I was in bondage. Drowning. And yet, I survived.

When you've hit the lowest point— when you've sunk as deep as shame can take you—if you can still lift your eyes, the only thing left to see… is *God.*

I asked Him to take away the thing in me that led me to infidelity. And He did.

I asked Him to refocus my heart. And He has.

I asked Him for peace. And He gives it, daily.

By Design was born from that place— from a man who has been shattered, reshaped, and redeemed. This isn't a book written by a theologian or a scientist—it's written by a father, a husband, a sinner, and a son. A man who once needed proof. A man who found it—not just in Scripture, but in science.

I've come to believe with all my heart that we are not random. The DNA that encodes life, the galaxies that spiral in order, the history that follows patterns, the human heart that longs for love—it all points to design. And the Designer? He knows your name.

This book is for those who want to believe, but struggle. For those who've been wounded. For those who think it's too late.

It's not too late.

5

You are not an accident.
You are not forsaken.
You are not unloved.

You were made on purpose.
You were made with love.
You were made... *by design*.

Dedication

To **God the Father** –
For Your mercy, Your patience, and
Your unshakable love.
Thank You for forgiving me when I
couldn't forgive myself,
for calling me Your son when I didn't
feel worthy to be claimed,
and for writing a story I couldn't see
while I was in the middle of it.

To **Jesus Christ** –
My Savior and Redeemer,
Thank You for Your sacrifice, for
loving me first,
and for showing me what real love
looks like—nails and thorns,
grace and blood, strength and
surrender.

To **the Holy Spirit** –
Thank You for speaking to me even
when I didn't listen,
for moving me when I was still,
and for guiding my feet back onto the
path of life.

And to **my wife** –
You are the salt and the light in my
life.
You held fast when I deserved to be
let go.

You saw the man I could be when I
was at my worst.
You are worth far more than rubies—
and I will spend the rest of my life
trying to be worthy of your
forgiveness,
your love, and your faith.

Chapter 1: DNA, Design, and Divine Insight

Introduction: DNA, Evolution, and the Framework for Design

Although I do not support evolution as the definitive explanation for life on Earth, I recognize its profound influence on modern scientific thought and refer to it in discussions where context requires it. Evolutionary theory has provided a prevailing framework for understanding biology over the past century. Yet, despite its widespread acceptance, it leaves profound questions unanswered—questions about the origin of information, the fine-tuning of life, and the uncanny harmony between the physical laws of the universe and the conditions necessary for life to exist.

In this chapter, we will explore the striking contrast between randomness and intentional programming. We will look closely at DNA, the digital language of biology, and examine how its complexity and elegance transcend chance. We will also consider the cosmic structure of the universe, the sequence of events recorded in Genesis, and the extraordinary parallels between ancient Scripture and modern science. Through the lens of biology, cosmology, and information theory, we will uncover a consistent and resounding theme: not disorder, but design. Not chaos, but code. Not accident, but architecture.

In the hush of a dimly lit lab, a biologist leans back from her microscope, staring at the double helix sequence glowing on the monitor. After hours of decoding and analyzing, a single thought echoes louder than her data points: *This looks written.* Her mind drifts—not to randomness or mutations, but to something ancient. *"In the beginning..."* she recalls the phrase from her childhood. Could the code she's reading be more than chemistry? Could it be a signature?

The Code Analogy: DNA and Video Games

Imagine this: a 12-year-old sits in a dark room, surrounded by flickering screens. His fingers dance across the controller, piloting a character through a world of impossible beauty— sapphire skies, molten caverns, neon cities. He's not amazed that the world exists. He's amazed that someone made it.

Video games are code. Code is design. Design requires a designer.

Now zoom in—not on the screen, but into your body. Inside your cells are strands of DNA—a biological language written in four chemical letters. This code doesn't just simulate life; it creates it. It's a blueprint, a program, a digital instruction set. It tells your cells how to grow, how to repair, even how to die.

It's not chemistry. It's software. And every living thing runs on it.

Just as a game cannot write itself, DNA cannot arise from randomness. The more we learn, the more we realize—we're not glitches in the system. We're intentional.

"In the beginning, God created..." That's not just theology. It's technological. It's mathematical. It's biological. And it's deeply personal.

If DNA is a message, who sent it?

DNA is not merely a string of chemicals—it is a language. It contains syntax, grammar, and rules of interpretation. Functionally, it behaves much like a sophisticated programming language, a structured, hierarchical system of instructions that governs the form and function of every living organism. Encoded within its four-letter alphabet—adenine (A), thymine (T), cytosine (C), and guanine (G)—is the blueprint for every protein, every trait, every living cell.

When scientists compare genomes across species, they often find shared sequences, or homologous regions. Some interpret this as evidence for common ancestry, as proposed in Darwinian evolution. The logic goes: if humans and chimpanzees share over 98% of their DNA, we must have descended from a common ancestor. But this interpretation hinges on a key assumption—that similarity necessarily means shared lineage. Let's consider a more insightful analogy.

Take *Pong* (1972), one of the earliest arcade games, and *Super Mario Bros.* (1985), a landmark title that helped define a generation of gaming. These two games could not be more different in graphics, gameplay, or complexity. Yet, both were written in 6502 Assembly language—a low-level programming language used for early game development. Despite their differences, they share key coding structures:

- **Game loops**: Both games rely on continuous input-output cycles to run, akin to metabolic feedback loops that keep a cell or organism alive.
- **Input handling**: Just as video games respond to user controls, DNA responds to environmental triggers—temperature, light, chemical signals—that activate or suppress genes.
- **Collision detection**: Video games calculate interactions

between objects on screen, much like cells detect and respond to external forces, infections, or injuries.

Now, no one claims that *Super Mario* evolved from *Pong*. They are clearly distinct works, designed with intentionality. Yet both games reflect the fingerprints of intelligent programmers who reused efficient code across platforms.

Likewise, shared genetic code does not necessarily point to descent. It may instead point to a designer reusing optimal patterns, reapplying successful frameworks, and embedding universal principles within different forms of life. The logic of similarity through design is as powerful as similarity through ancestry—perhaps more so, because it acknowledges engineering intelligence rather than attributing complex function to unguided mutation.

Genetic Programming and the Case for Design

When we examine DNA through the lens of computer science or systems engineering, what emerges is a marvel of efficiency and foresight. Shared genetic sequences among different organisms may reflect:

- **A universal design framework**, similar to how game developers use shared engines or code libraries across different games. This explains both conservation and variation within the system.
- **Code optimization**, where reliable patterns—like start codons, promoter regions, or exon-intron architecture—are reused for functionality, just as programmers reuse tried-and-true subroutines.
- **Functional architecture**, allowing biological systems to operate within defined parameters while retaining adaptability. Genes may be turned on or off in response to internal and external signals,

ensuring dynamic resilience
without random drift.[1]

Biological similarity, in this view, does
not require evolutionary lineage.
Instead, it becomes evidence of
rational design—a single intelligent
source applying consistent principles
across varied life forms. Just as
multiple applications can be written in
the same programming language
without being derivative of one
another, organisms can share code
without sharing ancestry.[2]

The sheer elegance of this system
defies reduction to randomness. It is
not an accident that genes follow
syntax and logic. It is not a
coincidence that living systems display
redundancy, error correction, and
adaptive mechanisms. These are

---

[1] Sanford, John C. Genetic Entropy and
the Mystery of the Genome. Lima, NY:
FMS Publications, 2005.

2 Meyer, Stephen C. Signature in the Cell:
DNA and the Evidence for Intelligent
Design. New York: HarperOne, 2009.

hallmarks of foresight, not flukes of time.

DNA vs. Artificial Intelligence

Artificial Intelligence (AI) is often portrayed as the pinnacle of human innovation—a self-learning system capable of mimicking certain aspects of cognition. But when compared to DNA, AI appears as a crude imitation, lacking the grace, scalability, and efficiency found in nature's own code.[3]

DNA outperforms AI in nearly every category:

- **Self-repairing**: DNA has built-in mechanisms to detect and correct errors— proofreading enzymes and mismatch repair systems that maintain genomic integrity.

---

3 Efrati, Shai, Amir Hadanny, Yair Fishlev Gabbay, and Yafit Berkovitz. "Hyperbaric oxygen therapy increases telomere length and decreases immunosenescence in isolated blood cells: A prospective trial." Aging 13, no. 15 (2020): 20935–20952.

AI, by contrast, is fragile. A small glitch can crash an entire system.

- **Self-replicating**: A single DNA molecule can direct the formation of an entire organism through cell division and differentiation. AI cannot reproduce itself; it must be manually duplicated and maintained.

- **Dual-coded**: DNA contains instructions for both structure (hardware) and operation (software). It builds the body and governs its processes. AI separates code from form—it cannot create physical machines by itself.

- **Power-efficient**: DNA operates trillions of processes within the human body using minimal energy. AI requires immense data centers, consuming vast amounts of electricity to simulate narrow tasks like language translation or image recognition.

One researcher captured this contrast with a striking quip:
"DNA performs in 700MB what AI

needs 800GB to attempt—and still fails to match."

If AI is a sleek modern app—clever, but limited—DNA is the operating system of life. It is foundational, not optional. It governs from the inside out, scaling from a single fertilized cell to the human brain, with all its neurons, memories, and consciousness. And it does so without crashing.

Such sophistication cannot be the result of trial and error. It reflects precision beyond our current technology—let alone what random processes could produce over time.

Genesis and the Origin of Light

Genesis 1:1–3 reads:

> *"In the beginning, God created the heavens and the earth. And the earth was without form, and void; and darkness was upon the face of the deep. And the Spirit of God moved upon the face of the waters. And God said, 'Let there be light,' and there was light."*

This passage, written millennia ago, reflects uncanny alignment with modern cosmology. First, there is a **beginning**—a singular event that marks the origin of time, space, and matter. This aligns with the Big Bang theory, which affirms the universe had a definitive starting point.

Then, there is **primordial darkness**, an opaque cosmos filled with hot, dense plasma—exactly what physics describes before the release of light.

Finally, light appears by command. In scientific terms, this moment reflects what physicists call the **recombination era**, which occurred approximately **380,000 years after the Big Bang**. Before this time, the universe was filled with a hot, dense plasma of photons, electrons, and protons, which scattered light in all directions, making the universe opaque. But as the universe expanded and cooled, it reached a critical temperature ($\sim 3{,}000$ K) that allowed electrons to combine with protons to form neutral hydrogen atoms. This process, known as **recombination**, dramatically reduced the scattering of photons. As a result, **light decoupled from matter** and was free to travel through space. This event is referred to as **"photon decoupling."**[4]

The light released at that moment still travels through the cosmos today and is detectable as the **Cosmic Microwave Background (CMB)**—a faint, nearly uniform glow of microwave radiation that permeates the entire sky. Discovered in 1965 by Arno Penzias and Robert Wilson, the CMB provides one of the most compelling confirmations of the universe's early conditions. Its existence is a direct echo of the moment when **"God said, 'Let there be light'" (Genesis 1:3)**—a profound convergence of theological declaration and cosmological evidence. This divine act of illumination, encoded into the very structure of space-time, continues to whisper across the cosmos in the form of these ancient photons.

When Nobel laureate George Smoot analyzed this radiation, he remarked that observing these patterns was like

---

[4] Penzias, Arno A., and Robert W. Wilson. "A Measurement of Excess Antenna Temperature at 4080 Mc/s." The Astrophysical Journal 142 (1965): 419–421.

"seeing the face of God." A poetic, yet deeply fitting statement for a universe that speaks in echoes of its creation.

The Separation of Light and Darkness

Genesis 1:4–5 continues:

> *"And God saw the light, that it was good: and God divided the light from the darkness. And God called the light Day, and the darkness he called Night."*

In cosmological terms, this parallels the end of the **Cosmic Dark Ages**— a time when, after the Big Bang, the universe expanded and cooled, but light from the first stars had not yet formed. Then, as the first stars ignited, photons flooded the cosmos, ionizing hydrogen and marking the dawn of visible structure.

Only with light can the concepts of "day" and "night" become meaningful. Light defines time by its presence and absence. Genesis captures this fundamental threshold— when illumination became a property of the universe, and structure emerged from shadow.

The Firmament and Ancient Cosmology

Genesis 1:6–8 speaks of a "firmament" dividing waters above from waters below. While ancient cultures envisioned this as a solid dome overhead, the functional reality described maps closely onto Earth's **atmosphere**:

- **The waters below** refer to the seas, rivers, and subterranean reservoirs.
- **The waters above** describe vapor layers, clouds, and atmospheric moisture suspended in the sky.
- **The firmament** is the atmospheric barrier— transparent yet protective— that separates these two realms and enables weather, climate, and the water cycle.

Though framed poetically, this separation is essential for life. Earth's atmosphere filters radiation, maintains temperature, and distributes water—

just as Genesis describes a structured
expanse, stretched out to sustain life.

Dry Ground, Vegetation, and Earth's Oxygenation

Genesis 1:9–13 records

> *"And God said, 'Let the waters under
> the heavens be gathered together into
> one place, and let the dry land appear,'
> and it was so; and God called the dry
> land Earth, and the gathered waters
> He called Seas, and God saw that it
> was good, and God said, 'Let the earth
> sprout vegetation: plants yielding seed
> and fruit trees bearing fruit in which is
> their seed, each according to its kind, on
> the earth,' and it was so, and the earth
> brought forth vegetation: plants yielding
> seed according to their own kinds, and
> trees bearing fruit in which is their seed,
> each according to its kind, and God
> saw that it was good, and there was
> evening and there was morning, the
> third day."*

In Genesis 1:9–13, we find a depiction
of Earth's transformation as dry
ground emerges from the waters and
plant life takes root. The process
described aligns intriguingly with

scientific insights into Earth's early history.

Geologists have found that, during its initial phases, Earth was primarily covered by vast oceans. Tectonic forces, along with the accumulation of sediment, gradually raised portions of the seafloor, giving rise to continents and islands. This would correspond to the "gathering together of the waters" mentioned in the passage, creating distinct areas of land and sea.[5]

The emergence of dry land set the stage for the development of primitive plant life. Long before complex vegetation existed, microscopic organisms, especially cyanobacteria, began harnessing sunlight to perform photosynthesis. This biological innovation triggered the Great Oxygenation Event about 2.4 billion years ago—a landmark period when oxygen first began accumulating in Earth's atmosphere. This atmospheric shift not only made the air breathable for future complex life forms, but it

---

[5] Footnote: Holland, Heinrich D. "The oxygenation of the atmosphere and oceans." Philosophical Transactions of the Royal Society B: Biological Sciences 361, no. 1470 (2006): 903–915.

also enabled the proliferation of diverse plant species. Over millions of years, simple algae and diatoms contributed heavily to oxygen production, followed by mosses, ferns, and eventually flowering plants. These developments laid a foundation for ecosystems, regulating carbon cycles, enriching soils, and maintaining the planet's habitability.

Notably, Genesis places the creation of plant life before that of animals. This sequence corresponds with evidence from paleobiology, which shows that complex plants and oxygen-rich conditions emerged before the explosion of animal diversity. While the Bible uses theological language rather than scientific terminology, the order presented aligns with what modern science has revealed: the establishment of stable land, the rise of photosynthetic life, and the subsequent preparation of Earth's environment for more complex organisms. This harmony of scriptural narrative and scientific observation suggests a remarkable parallel between the ancient text and our understanding of Earth's biological and geological history.

Celestial Bodies and Atmospheric Clarity

Genesis 1:14–19 describes the appearance of the sun, moon, and stars—not as gods to be worshiped, but as **instruments of time**: "for signs, and for seasons, and for days, and years."

Stars existed before Earth, but their visibility was likely obscured by Earth's early atmosphere—thick with volcanic gases and particulate matter. As photosynthetic life transformed the air, the sky cleared, and the **heavenly lights** became visible markers for navigation, calendars, and agricultural cycles.[6]

Genesis emphasizes function over form, describing not the creation of celestial bodies themselves, but their **purpose** to serve, not to rule. This stands in stark contrast to pagan mythologies that deified the sun and

---

[6] Footnote: Catling, David C. Atmospheric Evolution on Inhabited and Lifeless Worlds. Cambridge: Cambridge University Press, 2017.

moon. Genesis deconstructs idolatry and affirms divine order.

Aquatic Life and Flying Creatures

Genesis 1:20–23 speaks of marine life and birds arising from the waters.

Science affirms this progression. The first multicellular organisms arose in oceans—sponges, jellyfish, and eventually complex vertebrates. Over time, evolutionary theory proposes that some dinosaurs developed feathers and took flight. Fossils like **Archaeopteryx** show characteristics of both reptiles and birds.

While interpretations vary, the sequence in Genesis remains remarkably accurate: sea life first, then avian life—a reflection of real natural history encoded in ancient text.

Land Animals and Human Uniqueness

Genesis 1:24–31 ends with the creation of land animals and, finally, humans.

Modern biology confirms that **tetrapods** gave rise to mammals, and eventually to **Homo sapiens**. But Genesis goes further than anatomy. It claims that humans are made **in the image of God**—a theological truth that transcends biology.

Humans alone:

- **Create civilization** with architecture, writing, and governance.
- **Invent abstraction**, from mathematics to philosophy.
- **Build moral systems** based on justice and compassion.
- **Worship and wonder**, contemplating origins, purpose, and eternity.

Science tells us what we are. Genesis tells us **who we are**, and why we matter.

The Genius of the Sequence

Genesis 1 presents a strikingly accurate unfolding of creation:

1. Cosmic origin – 1:1
2. Light emerges – 1:3
3. Sky and sea – 1:6–8
4. Land and vegetation – 1:9–13
5. Celestial visibility – 1:14–19
6. Aquatic and flying life – 1:20–23
7. Land animals and humans – 1:24–31[7]

For a text written over 3,000 years ago, this sequence displays stunning insight. Which raises the ultimate question:

How did Moses know this?

---

[7] Footnote: Ross, Hugh. The Genesis Question: Scientific Advances and the Accuracy of Genesis. Colorado Springs: NavPress, 2001.

Code, Compilers, and the Mystery of Execution

Just as software must be compiled and executed, so too must DNA. In computing:

- **Code is written** → interpreted by a compiler → run by a processor.[8]

In biology:

- **DNA is the code.**
- **RNA acts as the compiler**, transcribing the genetic script.
- **Ribosomes are the processors**, translating instructions into proteins.
- **Proteins are the result**—the physical expression of life.

It's a perfect pipeline. But such a system cannot arise gradually. It must exist **whole**, or not at all. Without DNA, no RNA. Without RNA, no proteins. Without proteins, no life.

---

[8] Footnote: Behe, Michael J. Darwin's Black Box: The Biochemical Challenge to Evolution. New York: Free Press, 1996.

Who wrote this code?

DNA as a Divine Algorithm

DNA is not just a data structure—it is a **reality-making machine**. It:

- Adapts and repairs
- Builds complexity from simplicity
- Encodes memory and behavior[9]

From a single microscopic seed grows a towering oak. From a fertilized egg, a conscious human being. And yet, the building blocks—atoms and molecules—possess no will, no knowledge. They are merely the medium. The **message** lies in the code.

This is no random arrangement. It is a script. A signal. A **message from the Mind behind matter**.

---

[9] Footnote: Gitt, Werner. In the Beginning Was Information. Bielefeld, Germany: Christliche Literatur-Verbreitung, 2006.

Conclusion: Life by Design

Genesis opens with a voice:
"And God said…"
DNA begins with a code:
"ATG…"
Both are acts of speech. Both
command existence. Both create life.
From the vastness of galaxies to the
elegance of the genome, design is
everywhere—layered, logical, and
luminous. Each star that blazes and
every cell that divides speaks of
intention, of a thought brought to
fruition. Science explains the
mechanisms—gravity's pull on the
heavens, chemical bonds forming the
building blocks of biology. But only
design explains the meaning. Only
design gives these mechanisms their
purpose, transforming what would be
mere processes into the intricate
tapestry of life.

Life did not arise from chaos.
It was composed, like a symphony
from a masterful composer.
Not guessed.
But given, as a gift handed down
through the ages.
Not random.
But written—by design, its beauty as

evident in the smallest strand of DNA
as in the farthest edge of the cosmos.

Chapter 2: Genesis – Creation, Cosmology, and the Fall

Genesis 1:1–3 – The Universe Begins

"In the beginning God created the heavens and the earth… And God said, 'Let there be light,' and there was light."
— *Genesis 1:1, 3*

Genesis opens with a declaration so sweeping and profound that its implications extend into the very fabric of physics, cosmology, and philosophy. "In the beginning…" signals more than the start of a story—it announces the absolute origin of time itself. Before that moment, there was no sequence, no ticking clock, no before or after. Time had not yet begun to tick. The text then moves seamlessly to "God created the heavens," describing the creation of space, the vast framework in which all galaxies, stars, and planets would reside. Finally, "and the earth" introduces matter—the tangible, physical substance of the universe.

This singular verse, written over three thousand years ago, astonishingly parallels the foundational triad recognized by modern physics: **time, space, and matter**. And not only that they exist—but that they **originated together**, in a single, unified moment. This mirrors the essence of the **Big Bang Theory**, which posits that the universe began approximately 13.8 billion years ago from an infinitesimal point of extreme density and temperature—what physicists call a **singularity**.

As the singularity expanded, so did time and space, woven together in what Einstein later described as **spacetime**—a unified continuum that undergirds all physical reality. Matter and energy burst forth in a cosmic event that still echoes today. What Genesis describes as a divine act, science has only recently begun to model and measure.

Verse 2 adds a haunting and beautiful layer:

"Now the earth was formless and void, and darkness was over the surface of the deep…"

This aligns remarkably with what physicists describe as the universe's earliest stage—a turbulent, high-energy plasma state, opaque to light and without solid form. For nearly 380,000 years after the Big Bang, the universe was so hot and dense that photons—the particles of light—were trapped in a fog of charged particles. It was only after the universe cooled enough for electrons and protons to combine into neutral hydrogen atoms that light could move freely. This moment, called **recombination**, allowed light to decouple from matter and spread across the cosmos.

Today, that ancient light still permeates space as the **Cosmic Microwave Background Radiation (CMBR)**—the faint afterglow of creation itself. Accidentally discovered in 1965 by physicists Arno Penzias and Robert Wilson, this discovery won them the Nobel Prize and provided stunning empirical support for a singular cosmic beginning. Stephen Hawking later referred to it as "the greatest discovery of 20th-century cosmology."

Genesis 1:3 continues:

> *"And God said, 'Let there be light,'*
> *and there was light."*

This doesn't refer to sunlight, which appears later in the creation narrative. It refers instead to the **emergence of light itself**—a phase transition in the structure of the universe, consistent with what modern astrophysics describes as the separation of radiation from matter.

Equally profound is the theological richness of verse 2:

> *"And the Spirit of God was hovering*
> *over the waters."*

This poetic image suggests an active, organizing presence—a divine intelligence preparing the universe for order and life. In physics, the early universe emerged from a remarkably **low entropy state**—an incredibly precise configuration that allowed galaxies, stars, and ultimately life to form. This is an unsolved mystery: Why did the universe begin with such exquisite order when chaos was vastly more probable? Some physicists invoke the **anthropic principle**, suggesting we observe this order because we are here to do so—but this only reframes the question.

Even quantum theories, such as those involving vacuum energy or **zero-point fields**, still rely on fine-tuned parameters. The concept of a "hovering" divine presence aligns poetically with the delicate balance scientists observe at the foundation of the cosmos—what cosmologist Paul Davies calls the "Goldilocks Enigma."

Light, in Scripture and in science, marks a beginning. It is the **first measurable change**, the first separation, the first act of revelation. It is a physical phenomenon—and a spiritual metaphor. Genesis does not merely narrate the formation of matter; it announces the emergence of meaning.

The universe began not with randomness, but with radiance. Not chaos, but clarity. It began—**by design**.

Genesis 1:4–5 – Day and Night Established

> *"God saw that the light was good. And God separated the light from the darkness. God called the light Day, and the darkness he called Night."*
> — *Genesis 1:4–5*

The separation of light from darkness may appear simple, even poetic, but it reflects a process of cosmic and biological importance. Earth's rotation on its axis—established early in planetary formation—results in a recurring alternation between light and shadow. One hemisphere faces illumination while the other turns away, producing the first rhythm of terrestrial life: **day and night**.

Even before the sun was visible from Earth's surface, this cycle existed. The rotating Earth, even under a thick, opaque atmosphere, would have experienced alternating periods of brightness and darkness. As the atmosphere evolved and cleared, these transitions became sharper and more pronounced.

This rhythm of light and darkness profoundly shaped life on Earth. Virtually all organisms, from cyanobacteria to humans, follow **circadian rhythms**—biological clocks regulated by the 24-hour cycle. In humans, circadian rhythms affect sleep, hormone secretion, metabolism, immune response, and cognitive function. Disruption of this rhythm, such as from prolonged artificial lighting or shift work, has been linked to increased risks of cancer, heart disease, and neurological disorders.

Plants, too, are deeply tuned to this cycle. Through **photoperiodism**, they sense changes in day length to regulate flowering, dormancy, and germination. Migratory animals navigate vast distances based on daylight cues. Tidal and lunar rhythms govern behaviors in marine life.

The text also notes that "God called the light Day..."—introducing not just physical change, but **semantic order**. The act of naming in Genesis is never arbitrary. It signals **authority**, **definition**, and **function**. God doesn't merely create—He assigns purpose.

Scientifically, this corresponds to Earth's increasing atmospheric clarity. As oxygenation from plant life rose (see Day 3), the previously thick, hazy skies began to clear. Light, once scattered and dim, became directional and more distinct—allowing visual contrast between day and night.

Theologically, light has always symbolized knowledge, presence, and holiness. Darkness represents confusion, chaos, or the unknown. The separation of the two is the first act of **moral and cosmic order**. It is a template for the structure of creation—and the structure of life.

Genesis 1:6–8 – The Firmament and the Formation of the Atmosphere

"And God said, 'Let there be an expanse in the midst of the waters, and let it separate the waters from the waters.' And God made the expanse and separated the waters that were under the expanse from the waters that were above the expanse. And it was so. And God called the expanse

Heaven."
— Genesis 1:6–8

The Hebrew term *raqia*, often translated as "firmament" or "expanse," has long been debated. Ancient cultures often envisioned the sky as a dome—a solid canopy holding up celestial waters. But when viewed through the lens of modern science, Genesis 1:6–8 appears to poetically describe the **emergence of Earth's atmosphere**—a complex, dynamic structure crucial for sustaining life.

In the planet's early history, violent **volcanic activity** released vast amounts of gas from Earth's interior. Carbon dioxide, methane, ammonia, sulfur compounds, and water vapor filled the primitive atmosphere, creating a thick, hot greenhouse blanket. As the planet cooled, this water vapor condensed, forming massive clouds and eventually triggering torrential rains that fell for thousands of years—helping to form Earth's oceans.

This process naturally divided **"the waters below"**—the surface seas—from **"the waters above"**—

atmospheric vapor and clouds. Between them stretched the *raqia*, the atmospheric expanse that makes life possible.

This "expanse" is not empty—it is structured, layered, and designed with function in mind. Earth's atmosphere is composed of several distinct strata:

- The **troposphere**, where weather and the water cycle occur
- The **stratosphere**, which contains the ozone layer that shields life from ultraviolet radiation
- The **mesosphere**, **thermosphere**, and **exosphere**, which transition into outer space

Each of these layers plays a specific role in filtering solar radiation, regulating temperature, and sustaining life. The formation of the atmosphere was a pivotal moment—without it, the Earth would be lifeless, scorched by cosmic rays or frozen by the vacuum of space.

The text also says, "God called the expanse Heaven." In Hebrew, *shamayim* is a plural word that encompasses both **the sky** and **the heavens beyond**. This dual meaning captures Earth's place as a boundary point between terrestrial waters and the celestial realms. The poetic phrasing contains a scientific reality: Earth's atmosphere sits delicately between ocean depths and infinite space—thin, protective, and essential.

Today, astrobiologists confirm that a planet's atmosphere is one of the most critical factors in making it **habitable**. Too thick, and temperatures soar. Too thin, and radiation sterilizes the surface. Earth's atmosphere is not just adequate—it is astonishingly fine-tuned. Genesis speaks of division, separation, and naming, not as myth, but as layers of **intentional structure**—delicately calibrated and **by design**.

Genesis 1:9–13 – Dry Land and the Birth of Vegetation

> *"And God said, 'Let the waters under the heavens be gathered together into one place, and let the dry land appear.'… And God said, 'Let the earth sprout vegetation, plants yielding*

Here we witness two profound transitions: the geological emergence of **dry land** and the biological emergence of **plant life**.

Geologically, Earth's surface was once a chaotic mix of molten rock and global oceans. Over time, tectonic activity caused crustal plates to rise, collide, and stabilize—lifting portions of land above the water. The ancient continents, such as Vaalbara and later Rodinia, were the result of these powerful forces. The **"gathering of waters"** into ocean basins and the **"appearance of dry land"** mirror what geologists observe as the slow sculpting of Earth's surface.

Then comes one of the most significant biological events in Earth's history:

"Let the earth sprout vegetation…"

At first glance, this may seem like a poetic flourish, but from a scientific standpoint, it reflects a pivotal transformation in the planet's biosphere. Early Earth's atmosphere lacked **free oxygen**, making it inhospitable to animal life. But the appearance of **photosynthetic organisms**, like cyanobacteria, began a slow but profound process. Through photosynthesis, these microorganisms took in sunlight, carbon dioxide, and water—and released oxygen as a byproduct.

This led to what scientists call the **Great Oxygenation Event**, approximately 2.4 billion years ago, when oxygen began to accumulate in the atmosphere. This event dramatically changed Earth's chemistry and paved the way for complex aerobic life.

After cyanobacteria, algae, mosses, and simple vascular plants followed. These primitive plants not only generated more oxygen but also helped stabilize soil, regulate climate, and form the basis of the global food chain. From this point, Earth became increasingly green, resilient, and habitable.

Genesis notes that plants were **"yielding seed according to their kinds,"** highlighting an important feature of biological systems:

reproductive fidelity within species. While evolution emphasizes variation and speciation over time, Genesis emphasizes stability and bounded variation—each kind reproducing after its own kind, with built-in limits to change. This concept resonates with the modern understanding of **genetic inheritance**, where variation exists but does not lead to infinite transformation without boundaries.

Plant life was not only biologically important—it was **foundational**. Plants fixed carbon, produced oxygen, stabilized ecosystems, and fed future life. The Genesis account places their creation before animals—exactly as the fossil record affirms. And like everything in the Genesis narrative, their appearance follows a logical sequence—structured, necessary, and deeply **intentional**.

Genesis 1:14–19 – Lights in the Sky: Celestial Order and Visibility

"And God said, 'Let there be lights in the expanse of the heavens to separate the day from the night… and let them be for signs and for seasons, and for

days and years.'"
— *Genesis 1:14*

This passage often generates confusion unless one understands that Genesis is written from the **perspective of an observer on Earth**, not from outer space. Genesis 1:14 does not claim that the sun, moon, and stars were created at this point in absolute terms—rather, they become **visible and functionally significant** to Earth's surface.

Earlier in Genesis 1:1, it already declared, "In the beginning, God created the heavens and the earth," implying that the celestial bodies already existed. The emphasis in verses 14–19 is not origin, but **function and visibility**: the sun, moon, and stars becoming distinguishable and serving a **purpose** in the created order.

This change likely corresponds to a pivotal transformation in Earth's atmosphere. As plant life proliferated (see Day 3), oxygen began to rise, and methane levels dropped. Over time, this caused the once-thick, cloudy atmosphere to become more **transparent**—what scientists call the

atmospheric transparency window. As the air cleared, Earth's rotation allowed the sun to mark days, the moon to reflect light at night, and the stars to appear regularly in the sky.

Genesis emphasizes the **purpose** of these bodies:

- **To divide light from darkness**
- **To mark time—days, seasons, and years**
- **To serve as signs**

These functions are still critical today. Earth's **axial tilt (23.5°)** and elliptical orbit around the sun create **seasonal changes**, while the **lunar cycle** affects tides, agricultural planning, and even animal behavior. Ancient civilizations used the stars to navigate vast oceans and track the passage of time—confirming that these lights were not deities to be worshiped, as in pagan cultures, but **tools** within creation.

Astronomers like Guillermo Gonzalez have noted that Earth is not just ideally placed for life—but also uniquely situated for **cosmic discovery**. Our location in the galaxy, our clear atmosphere, and the size and position of our moon allow us to observe eclipses, cosmic events, and planetary motion with unusual precision. Genesis frames these observations as **designed**—to serve, inform, and point to deeper truths.

Psalm 19:1 declares,

> *"The heavens declare the glory of God, and the sky above proclaims his handiwork."*

Genesis 1:14–19 affirms this by framing the sky not just as a backdrop—but as a **calendar**, a **clock**, and a **canvas** of divine glory.

Genesis 1:20–23 – Life in the Seas and Sky

> *"And God said, 'Let the waters swarm with swarms of living creatures, and let birds fly above the earth…'"*
> — *Genesis 1:20*

This portion of Genesis marks the introduction of **animal life**—beginning with marine creatures and flying birds. Once again, the order is scientifically accurate: **life began in the sea** and **flying creatures arose later**, mirroring the fossil record.

The earliest known life forms appeared in the oceans. Fossils of **stromatolites**, built by cyanobacteria, date back over 3.5 billion years. From these microbial communities, marine ecosystems blossomed into diverse forms—jellyfish, mollusks, trilobites, and eventually fish with internal skeletons.

Birds, by contrast, appeared much later. According to paleontology, many scientists believe birds evolved from small theropod dinosaurs. Fossils like **Archaeopteryx**, discovered in the 19th century, possess both reptilian and avian traits—feathers, wings, and claws. These forms illustrate **structural continuity** that may reflect either evolutionary progression or an underlying design pattern, where similar genetic or morphological frameworks are reused across species.

Genesis describes each as created **"according to their kinds."** This phrase, repeated throughout the creation narrative, suggests **variation**

52

within boundaries. It allows for diversity, adaptation, and specialization—but resists the idea of one kind transforming into a wholly different one over time. This concept resonates with modern genetics, where species reproduce within fixed gene pools, allowing for microevolution (change within species) but offering no clear mechanism for macroevolution without design.

Ecosystemically, both marine life and flying creatures serve key roles. **Plankton and algae** remain vital to Earth's oxygen cycle, accounting for more than half the planet's oxygen production. Birds pollinate, disperse seeds, control insect populations, and migrate in tune with planetary cycles.

These are not haphazard life forms—they are embedded in a **web of symbiotic function**. Each creature contributes to a larger balance, one that appears meticulously arranged rather than randomly assembled.

Genesis 1:24–28 – Land Animals and the Image of God

> *"And God said, 'Let the earth bring forth living creatures...' Then God said, 'Let us make man in our image, after our likeness...'"*
> — Genesis 1:24, 26

The sixth day of creation brings the narrative to its climax. First, **land animals** are introduced—beasts of the earth, livestock, and creeping things. Then, with special emphasis, God creates **humanity**—not merely as another organism, but as the **image-bearer** of God.

Science again confirms the basic sequence. Fossil evidence shows that amphibians, reptiles, and mammals emerged long before Homo sapiens. Each group filled ecological niches and developed sophisticated biological systems—hearing, eyesight, warm-bloodedness, and social behaviors.

But the arrival of **humans** is categorically distinct. Unlike any other species, we are not just intelligent—we are **self-aware**, moral, creative, and spiritual. We build cities. We write poetry. We ponder eternity.

Genesis calls this being *adam*—man—made in the *tselem Elohim*, the **image of God**. This implies not physical appearance, but spiritual resemblance: rationality, relational capacity, moral awareness, and dominion. Only humans are commanded to **steward** creation. Only humans are given speech, naming rights, and a mission.

Modern science still struggles to explain **consciousness**. As Nobel laureate Roger Penrose and philosopher Thomas Nagel have noted, human awareness cannot be fully reduced to neurochemical activity. There is a **qualitative leap**—a mystery at the core of our being.

Genesis captures this by affirming that mankind is both **formed from the earth** and **breathed into by God** (Genesis 2). We are material and spiritual, temporal and eternal. And our identity, according to Scripture, is not a fluke of chance—but a direct expression of **divine intention**.

Genesis 1:29–31 – A Plant-Based Diet and Nutritional Design

> *"Behold, I have given you every plant yielding seed... You shall have them for food."*
> — *Genesis 1:29*

The creation account ends with provision. God declares that mankind—and all animals—are to eat from the abundance of the plant kingdom. This reflects a **pre-Fall** state of harmony: no predation, no violence, no bloodshed. All life was sustained through peaceful, photosynthetically derived energy.

Remarkably, modern nutritional science supports the benefits of a **plant-based diet**. Studies have consistently shown that diets rich in whole grains, fruits, legumes, and vegetables reduce the risk of:

- Cardiovascular disease
- Type 2 diabetes
- Certain cancers
- Obesity and inflammation

One of the most comprehensive studies, the **Adventist Health Study**, tracked vegetarians and vegans in Loma Linda, California—a population known for longevity and low chronic disease rates. The findings affirm that a well-balanced plant-based diet not only promotes health but may extend lifespan.

From an ecological standpoint, plant-based nutrition is also more **sustainable**. It requires fewer natural resources, produces less greenhouse gas, and supports biodiversity by reducing habitat loss.

Genesis portrays this diet not as deprivation, but as **abundance**—a world in which nourishment is freely given from the ground. This reinforces a vision of Edenic peace, where life was designed for flourishing, not competition.

Verse 31 concludes:

"And God saw everything that He had made, and behold, it was very good."

Not just functional. Not just beautiful. But **very good**—whole, ordered, and harmonious.

Conclusion: The Pattern of Creation

Genesis 1 is not a primitive myth—it is a majestic tapestry of theology, science, and language. Each day reflects not randomness, but **rhythm**. Not chaos, but **choreography**. The sequence—cosmos, light, atmosphere, land, vegetation, celestial bodies, aquatic life, flying creatures, land animals, and finally humans—bears the fingerprint of **intelligence**.

Genesis anticipates what science has slowly uncovered: a universe with a beginning. A planet fine-tuned for life. An atmosphere layered for protection. A biosphere balanced in function. A human species unique in calling.

From the stars above to the DNA within, every line of code echoes one truth:

It was not random.
It was **written—by design.**

Chapter 3: Genesis 2 – The Garden, the Man, and the Mystery of Life

> *"Then the LORD God formed the man of dust from the ground and breathed into his nostrils the breath of life, and the man became a living creature."*
> — Genesis 2:7

Genesis 2 does not contradict the grand, sequential vision of Genesis 1—it magnifies it. Where Genesis 1 speaks with the voice of a cosmic architect, Genesis 2 speaks with the touch of a sculptor. The focus shifts from galaxies to gardens, from stars to soil, from the universal to the personal. We are invited into a closer look at the creation of humanity—not as a footnote to the universe, but as its centerpiece.

This chapter is not a second creation account, but an expansion—an intimate window into the formation of man, the planting of Eden, the crafting of woman, and the covenant of marriage. It is the story of origins told not in scientific abstraction, but in personal relationship. Where Genesis 1 shows the design of the world, Genesis 2 shows the **designer's hand on the clay**.

Genesis 2:7 – Dust and Breath: The Chemistry of Human Life

> *"Then the LORD God formed the man of dust from the ground and breathed into his nostrils the breath of life…"*

This simple yet profound statement fuses two realities: **matter and spirit, earth and heaven**. Man is formed from the *adamah*—the ground—linking him forever to the physical world. The Hebrew words themselves reflect this connection: *adam* (man) and *adamah* (ground) share the same linguistic root, anchoring our identity to the dust beneath our feet.

Modern science confirms the elemental accuracy of this description. The human body is composed almost

61

entirely of elements found in Earth's crust: oxygen (65%), carbon (18%), hydrogen (10%), nitrogen (3%), calcium, phosphorus, potassium, and trace metals. These are the same elements that constitute soil, rock, and even stars. In that sense, Carl Sagan's famous quote—"We are made of star-stuff"—is not at odds with Genesis, but strangely aligned. Both affirm that we are **physically of the earth**.[10]

But Scripture adds what science cannot: **the breath of God**. The Hebrew word *neshamah* conveys far more than respiration. It is used elsewhere to describe the divine breath that imparts wisdom, understanding, and spiritual vitality (cf. Job 32:8). Man is not merely a biological machine. He is not animated only by neurons and synapses. He is a **living soul**, a conscious being with intellect, emotion, will, and moral capacity—because God breathed into him something that transcends physical law.

---

[10] Sagan, Carl. Cosmos. New York: Random House, 1980.

Biology still cannot account for this. Consciousness—the awareness of self and the ability to reason abstractly—remains a mystery. Neurologists can track brain activity, but they cannot explain **why we are aware** of being aware. No animal ponders its mortality. No chimpanzee writes poetry. No dolphin builds cathedrals or debates ethics. The **human spirit** remains singular in all creation.

Genesis answers why. Man is dust—and divine breath. He is a union of the temporal and the eternal, a being whose very composition reflects the mystery of life: not random matter, but **formed by purpose**, animated by spirit, and created **by design**.[11]

Genesis 2:8–14 – Eden: The Garden and Its Geography

> *"And the LORD God planted a garden in Eden, in the east… A river flowed out of Eden to water the garden, and there it divided and became four*

---

[11] Nagel, Thomas. Mind and Cosmos: Why the Materialist Neo-Darwinian Conception of Nature Is Almost Certainly False. New York: Oxford University Press, 2012

rivers..."
— *Genesis 2:8,10*

Unlike mythologies that place paradise in unreachable realms or celestial heavens, Genesis locates Eden in **real geography**. The garden is not a metaphorical space—it is grounded in identifiable rivers, lands, and mineral-rich regions. Genesis describes a single river flowing from Eden, dividing into four headwaters: the Pishon, Gihon, Tigris, and Euphrates.

Two of these—the **Tigris** and **Euphrates**—are still well known today. They define Mesopotamia, the "land between rivers," and were the cradle of ancient civilization. But the other two—**Pishon** and **Gihon**—have long been debated.

The Pishon and Wadi Al-Batin

Genesis 2:11 describes the Pishon as encircling the land of **Havilah**, a place rich in gold, bdellium, and onyx. This matches archaeological evidence from the **Arabian Peninsula**, particularly a now-dry river known as **Wadi Al-Batin**. Satellite imagery from NASA has revealed ancient river channels stretching through Kuwait and northeastern Saudi Arabia. These riverbeds were once flowing arteries of life, especially during the Holocene Wet Period (roughly 8000–3000 BC), when the region was green and habitable.

This wadi connects to the **Kuwait River**, a now-buried system that once emptied into the Persian Gulf. It would have joined the Tigris-Euphrates system, forming a complex and fertile hydrological network—matching the Genesis description of four rivers flowing from a common source.

The Gihon and the Karun (Kuron) River

The Gihon is said to encircle the land of **Cush**. While "Cush" is often associated with Ethiopia, historical references indicate that the name also applied to regions **east of Mesopotamia**, including parts of western **Iran**. The **Karun River**, the largest river in Iran, fits this location. It flows through the ancient land of Elam—rich in resources—and empties into the Shatt al-Arab, joining the Tigris and Euphrates near the Persian Gulf.

The convergence of these four rivers—Tigris, Euphrates, Karun, and the now-lost Kuwait River (Pishon)—centers around the **northern Persian Gulf**, near the borders of modern-day Iraq, Kuwait, and Iran. This area, once lush and densely settled, likely served as the historical foundation for the Garden of Eden.

The Decline of Eden's Fertility

Today, much of this region is desert. But Genesis never claims Eden was destroyed—only that **man was driven out**, and access was cut off (Genesis 3:24). Over time, the region's fertility vanished due to multiple factors:

- The **Holocene Drying Period** (~5000–3000 BC) led to a significant decline in rainfall, turning green pastures into barren desert.
- The **uplift of the Zagros Mountains** altered the flow of rivers, diminishing the Karun and causing massive sediment buildup.

- The **Kuwait River** dried up, and Wadi Al-Batin became a memory beneath layers of sand.
- The **rising Persian Gulf** submerged ancient delta regions, erasing much of the original landscape of Eden.

But the memory remained. Genesis preserves not just a theological truth, but a **geographical echo**—a garden once real, now hidden. The rivers tell the story. The land remembers. Eden was not a myth—it was a beginning.

Genesis 2:21–23 – The Rib and the Formation of Woman

> *"So the LORD God caused a deep sleep to fall upon the man, and while he slept took one of his ribs... and the rib that the LORD God had taken from the man He made into a woman."*
> — *Genesis 2:21–22*

This passage describes the creation of woman not in terms of dust or ground, but of **shared origin**. She is not fashioned from the soil as Adam was, but from his side—highlighting both her uniqueness and her oneness with the man. This is not just a narrative of anatomical construction; it is a theological proclamation of unity, intimacy, and purpose.

The Hebrew word translated "rib" is *tsela*, a term that elsewhere refers to the **side** of a structure, like the side of the Tabernacle (Exodus 26:20). This broader meaning suggests that what was taken from Adam was not merely a single bone but a **portion of his very being**—his structure, his substance, his frame. Woman is not an afterthought, but a counterpart: not made from his head to rule over him, nor from his foot to be beneath him, but from his side—equal, close to the heart, created to walk alongside.

This poetic and theological truth may also hold a fascinating biological parallel.

Rib Regeneration in Modern Science

For many years, it was believed that once removed, human bones could not regenerate in full. However, medical studies have shown that the

rib is an exception—a fact that aligns intriguingly with the Genesis account.

Research published in *Chest Surgery Clinics of North America* (2005) documented that if the **periosteum**—the membrane that covers the rib—is left intact, the rib can **completely regenerate** within 6 to 12 months. Surgeons have long harvested ribs for reconstructive procedures, particularly in facial and thoracic surgery, precisely because of this regenerative property. As long as the protective sheath remains, the rib will grow back—functionally, structurally, and biologically whole.[12]

This makes the rib unique among the major bones of the body and a fitting choice in Genesis 2. God takes from Adam something that is **life-giving, self-healing, and protected**—a part of his frame that will not leave him diminished.

---

[12] Kawai, Koichi, et al. "Rib regeneration after costectomy in humans: Histological and radiological studies." Interdisciplinary Cardiovascular and Thoracic Surgery 13, no. 3 (2011): 232–236.

The symbolism here is deep. The woman is built from living tissue, near the heart and under the arm—places of affection and protection. She is not described as an "other," but as a part returned. The Hebrew word used when God "made" the woman (*banah*) means "built" or "crafted"—a deliberate act of artistry. She is not an accident. She is a design.[13]

Genesis 2:23–25 – Covenant, Nakedness, and the First Marriage

> *"Then the man said, 'This at last is bone of my bones and flesh of my flesh; she shall be called Woman, because she was taken out of Man.'"*
> — *Genesis 2:23*

With this declaration, Adam speaks the first recorded words of any human being—and they come not as instruction or command, but as **poetry**. This is not a dry observation; it is a song of joy, recognition, and

---

[13] Guliuzza, Randy J. "Design Analysis Suggests that Human Reproduction is Engineered, Not Evolved." Acts & Facts 39, no. 4 (2010): 10–11

belonging. The man sees the woman and instinctively knows: *This is the one who completes me. This is my equal, my companion, my reflection.*

In calling her "Woman" (*ishah*), Adam uses a play on his own name (*ish*, man)—underscoring their shared identity and origin. She is not separate from him, but part of him. Their relationship is not hierarchical but relational—rooted in **unity, likeness, and mutual purpose**.

The next verse delivers the theological foundation of **marriage**:

> *"Therefore a man shall leave his father and his mother and hold fast to his wife, and they shall become one flesh."*
> — *Genesis 2:24*

This verse introduces a new kind of bond—stronger even than the familial ties of birth. The language of "leave and cleave" describes the act of **covenant commitment**. A man detaches from his first home to establish a new household—a union marked by faithfulness, trust, and physical, emotional, and spiritual intimacy.

This "one flesh" union is not just about reproduction; it's about **interdependence**, **vulnerability**, and **partnership**. It is the foundation for family, for society, and even for theology. Throughout Scripture, the imagery of marriage is used to describe God's relationship with His people:

- Israel is called God's bride (Isaiah 54:5)
- The Church is the Bride of Christ (Ephesians 5:25–27)
- Heaven culminates in a wedding feast (Revelation 19:7–9)

Just as Eve was taken from Adam's side, the Church comes from Christ—born of His sacrifice, cleansed by His blood, brought into union through grace.

> *"And the man and his wife were both naked and were not ashamed."*
> — *Genesis 2:25*

These final words of Genesis 2 are pregnant with meaning. In Eden, **nakedness** was not a cause for fear or shame—it was a symbol of **perfect trust, full acceptance**, and **complete**

transparency. There was nothing to hide. No sin. No exploitation. No insecurity.

This condition reflects the psychological and spiritual wholeness of unfallen humanity. The absence of shame points to a world untouched by guilt, trauma, or relational fracture. Man and woman were not only in harmony with each other, but also with themselves and with God. Their world was one of open communion— a peace that modern humanity still longs to recover.

But this verse is also a threshold. What begins as harmony will soon become **rupture**. The serpent waits. Temptation looms. The first lie will soon unravel the first love. Trust will fracture, shame will enter, and the gates of Eden will close behind humanity.

But for this moment, the scene is still perfect. A garden, a man, a woman, a Creator—and nothing between them. This is not a fairy tale. It is the first memory of mankind. A memory encoded in Scripture and echoed in the human heart.

This is where it all began. Not in abstraction or mythology, but in a real place, with real people, with real purpose—fashioned, formed, and **breathed into existence by design**.

Chapter 4: Genesis 3 – The Fall, the Fracture, and the Promise

"Now the serpent was more crafty than any other beast of the field that the LORD God had made."
— Genesis 3:1

Genesis 3 is the hinge on which the entire biblical narrative swings. In the first two chapters, we see a world bursting with goodness—light and life, order and purpose, peace and unity. God speaks, and creation flourishes. Man and woman live in perfect fellowship with their Creator and with one another. But here, in Genesis 3, a new voice enters. It is not the voice of life, but of subtle rebellion. Not the voice of truth, but of temptation dressed in reason. The serpent's words do not simply challenge God's command—they challenge God's very character.

With the serpent comes more than a conversation. He brings a distortion of reality, a reshaping of trust, and the seeds of sin. The story that began in light now enters the shadows—and nothing will ever be the same.

Genesis 3:1–5 – The Psychology of the Lie

"Did God actually say…?"

The serpent doesn't begin with an outright denial, but with a question. It is a rhetorical dagger veiled in curiosity. By asking, "Did God actually say, 'You shall not eat of any tree in the garden'?" the serpent twists God's generous provision into restrictive prohibition. He takes a command that was meant to protect and reframes it as a limitation designed to control.

This is the **psychology of deception**. The first lie does not reject truth outright—it reframes it. It plays on emotion, not evidence. The serpent's tactic mirrors what cognitive psychologists today call **cognitive distortion**: a mental filter that reshapes reality into something negative. Eve is not shown all the trees she can eat from—she is led to focus on the one she cannot.

The serpent isolates her from context, places her in dialogue with doubt, and subtly undermines her confidence in God's intentions. He minimizes the consequence ("You will not surely die") and magnifies the reward: "You will be like God, knowing good and evil." This is not merely temptation—it is an appeal to autonomy.

The fall begins not with appetite, but with **ambition**. The desire is not for fruit—it is to become **self-defined**, independent of divine guidance. The knowledge of good and evil, in this context, is not mere moral awareness, but the **claim to moral authority**. It is the human attempt to become the source of truth rather than its recipient.

At the core of sin is not action—it is a shift in **trust**. Eve does not fall because she is hungry. She falls because she begins to suspect that God is withholding something good. That suspicion fractures the relationship—and the act follows.

Genesis 3:6–7 – The Disobedience and the Death of Innocence

> *"She took of its fruit and ate, and she also gave some to her husband who was with her, and he ate."*

The act itself is deceptively simple. A gesture. A bite. But its implications are vast. With that first taste of disobedience, something cosmic fractures. The unity between man and God, between man and woman, and even within the human self, is shattered.

"Then the eyes of both were opened…" But this enlightenment brings no glory. Instead, it brings shame. For the first time, they see themselves not in wonder, but in **vulnerability**. They are exposed—physically, emotionally, spiritually. The same bodies that once inspired celebration now provoke discomfort. The knowledge they gained was not liberation—it was alienation.

They sew fig leaves together—not just to cover their bodies, but to cover their **loss of innocence**. This is the first act of self-justification, the first attempt to construct a barrier between shame and the truth. But no fig leaf—then or now—can cover the rupture of relationship.

Theologically, this moment introduces what the Bible later calls **spiritual death**: separation from the source of life. The trust that once bound God and humanity is severed. Physically, the long march toward biological death begins. The apostle Paul describes this as "the bondage of corruption" (Romans 8:21)—a creation subject to futility because of sin.

From a scientific angle, this is where **entropy** becomes a central experience of human life. The Second Law of Thermodynamics—the principle that systems tend toward disorder—mirrors what Genesis describes. While the law existed from the beginning, it is now tied to **human futility**. Death, decay, disease—these are not simply biological realities. Genesis frames them as the consequence of rebellion against the Giver of life.[14]

Genesis 3:8–13 – The Hiding and the Blame

> *"And they heard the sound of the LORD God walking in the garden... and the man and his wife hid themselves."*

This is one of the most poignant scenes in all of Scripture. The Creator walks among His creation, but the response is no longer joy—it is fear. Shame has entered the human heart. The instinct now is not to draw near, but to **withdraw**. This is the first moment of hiding—not just in body, but in spirit.

They hide from God. They hide from one another. And perhaps most tragically, they hide from themselves.

When God calls out, "Where are you?" it is not because He lacks information. It is a relational question,

---

[14] Atkins, Peter. The Second Law: Energy, Chaos, and Form. New York: Scientific American Library, 1984

not a geographical one. God is not locating their bodies—He is calling their hearts. The question is designed to **awaken their conscience**. It is an invitation to come out from hiding.

But instead of confession, there is **blame**:

- Adam blames Eve—and subtly blames God: "The woman whom *You* gave to be with me..."
- Eve blames the serpent: "The serpent deceived me..."[15]

This is the birth of what psychologists now call **defensive projection**—the shifting of guilt onto others. It is the refusal to take ownership. The honesty of Genesis is stark: human beings, even in their first failure, choose excuses over repentance.

And yet, God's response is not abandonment. He does not walk away. He stays. He engages. He pronounces judgment—but also makes a promise.

---

[15] Plantinga, Alvin. Warranted Christian Belief. New York: Oxford University Press, 2000.

Genesis 3:14–15 – The Curse and the First Promise (Protoevangelium)

"I will put enmity between you and the woman, and between your offspring and her offspring; he shall bruise your head, and you shall bruise his heel."
— Genesis 3:15

God begins His judgment not with the humans, but with the serpent. No questions are asked. No defense is offered. The sentence is swift and direct. The serpent is cursed—uniquely among the creatures. He is condemned to crawl, symbolically lowered, destined for humiliation and defeat.

But it is in verse 15 that we find the **first glimmer of hope**. This verse is known as the **Protoevangelium**—the "first gospel." Amid the wreckage of rebellion, God plants a seed of redemption.

- A descendant of the woman will rise.

- There will be conflict—
  between good and evil, truth
  and lies, life and death.
- This descendant will crush the
  serpent's head—a fatal,
  decisive blow.
- But in the process, the serpent
  will bruise His heel—a
  wounding, but not a victory.

This is more than a struggle between
humans and snakes. It is the **cosmic
battle** between the enemy of souls
and the coming Messiah. The promise
of a "seed" of the woman already
hints at something miraculous—seed
is typically attributed to the man.
Here, God plants the concept of a
virgin-born Savior, a Redeemer who
will undo what the serpent has done.

At Calvary, this prophecy finds its
fulfillment. Christ, born of a woman,
crushed the power of sin and death.
His heel was bruised—pierced by
nails, bloodied by suffering. But in
that very act, the serpent's head was
crushed. Death was defeated. The
curse was answered

Genesis 3:16–19 – Pain, Toil, and the Entrance of Entropy

> *"To the woman He said, 'I will surely multiply your pain in childbearing...'*
> *To Adam He said, 'Cursed is the ground because of you... By the sweat of your face you shall eat bread... till you return to the ground.'"*
> — *Genesis 3:16–19*

Sin is not without consequence, and yet even in judgment, God's response is precise, measured, and just. Each sentence corresponds to the area of dominion God had entrusted to Adam and Eve. The woman, who was designed to nurture life, will now experience **pain** in doing so. The man, placed in the garden to tend and cultivate, will now experience **toil**.

To the Woman – Pain and Relational Strain

Pain in childbirth is intensified—not only in the physical sense but also emotionally and spiritually. The miracle of bringing life into the world becomes marked by suffering—a reminder of the brokenness that sin introduced. This is not punishment for reproduction, but a symbol of the cost that now accompanies life and legacy in a fallen world.

The second part of the woman's judgment—"Your desire shall be for your husband, and he shall rule over you"—signals a disruption in relational harmony. The equal partnership once shared in Eden is now threatened by **conflict and control**. The word "desire" here may suggest a desire to control, paralleled in Genesis 4:7, where sin's "desire is for" Cain, yet he must master it. What was once a relationship of mutual submission is now marred by tension, power imbalance, and distorted roles—not because of God's design, but because of human sin.

To the Man – Toil, Frustration, and Decay

Adam's judgment focuses on the ground—his place of labor. What was once a source of delight is now a source of struggle. The earth will resist his efforts. Thorns and thistles will rise uninvited. Sweat will replace ease.

His days will be spent wrestling not only with the soil, but with **futility**.

This struggle is echoed in all human effort. Work becomes difficult. Goals are harder to reach. Even when fruit comes, it spoils, fades, or fails to satisfy. The ground itself, symbolic of the wider creation, is **cursed**—not with annihilation, but with disorder.

From a scientific perspective, this is again consistent with the **Second Law of Thermodynamics**. Entropy—the natural tendency toward disorder—is a fundamental principle of the physical world. But Genesis gives it a moral context. Entropy is no longer just a physical reality—it becomes a spiritual metaphor: that all things, apart from divine redemption, tend toward decay.

"You are dust, and to dust you shall return."

This solemn declaration affirms a biological truth—our bodies are composed of the same elements found in soil. But it also introduces a spiritual reality: that death is now our shared fate. We are mortal not by nature, but by consequence.

Paul reflects on this in Romans 5:12:

"Therefore, just as sin came into the world through one man, and death through sin, and so death spread to all men because all sinned…"

This moment in Eden becomes the origin point of **human mortality**, not just in body, but in spirit.

Genesis 3:20–24 – Exile and the Guarded Gate

> *"Therefore the LORD God sent him out from the garden of Eden… He drove out the man, and at the east of the garden of Eden He placed the cherubim… to guard the way to the tree of life."*
> — *Genesis 3:23–24*

Though Adam and Eve do not die immediately, their exile from Eden is a form of death—**separation** from the presence of God and from the Tree of Life. They lose access to the source of ongoing immortality. But this is not merely punitive—it is protective. God declares in Genesis 3:22 that if man were to eat from the Tree of Life in his fallen state, he would live forever in corruption. Thus, the exile is **both judgment and mercy**.

The placement of cherubim—heavenly guardians—at the east of Eden signals that **re-entry cannot be achieved by human effort**. The way back is not blocked arbitrarily; it is guarded because restoration requires atonement. The presence of a sword, flashing back and forth, reminds us that only by sacrifice, only by passing through judgment, can the way to life be reopened.

This exile is echoed throughout the Bible:

- Israel is exiled from the Promised Land.
- Humanity is alienated from God.
- The temple curtain separates man from the Holy of Holies.

But Scripture also promises **return**. Just as Eden was closed, so too is it prophesied that a Redeemer would reopen the path. Revelation 22 draws the story to completion—not just with restored paradise, but with access to the Tree of Life:

> *"Blessed are those who wash their robes, so that they may have the right to the tree of life and may enter the city by the gates."*
> — *Revelation 22:14*

What was lost in Genesis is regained in Revelation. The tree once forbidden is now freely offered—to those redeemed by the blood of the Lamb.

Conclusion: The Fracture of Creation and the Foreshadow of Redemption

Genesis 3 is the origin of every fracture we feel in our world. It explains the source of shame, suffering, broken relationships, violence, aging, toil, and death. But even as judgment is spoken, **hope is planted**.

God clothes Adam and Eve with garments of skin (Genesis 3:21)—a sign of compassion and the first **bloodshed** to cover sin. He speaks the Protoevangelium—the first promise of a Savior. He guards the Tree of Life not in finality, but until the time is right.

89

Science can describe entropy. It can measure death. But Genesis tells us **why** they entered our story—and more importantly, how they will one day be overcome. We are not broken by poor design. We are fractured by broken relationship. And the story of Scripture is the story of a Designer who does not abandon His creation, but enters into it—bleeds for it, redeems it, and restores it.[16]

The seed has come.
The heel has been bruised.
The serpent's head is crushed.[17]
The way back to Eden is open.

From the garden to the cross, and from the cross to the **new creation**— this is the arc of redemption.

And it begins, as all true stories do, **by design**

---

[16] Meyer, Stephen C. Return of the God Hypothesis: Three Scientific Discoveries That Reveal the Mind Behind the Universe. New York: HarperOne, 2021

[17] Craig, William Lane. Reasonable Faith: Christian Truth and Apologetics. Wheaton, IL: Crossway, 2008.

Chapter 5: Genesis 4 – Cain, Culture, and the Echoes of Eden

> *"Now Adam knew Eve his wife, and she conceived and bore Cain, saying, 'I have gotten a man with the help of the LORD.'"*
> — *Genesis 4:1*

Genesis 4 opens with hope. The first birth in human history follows the first exile. Adam and Eve, now driven from Eden, begin a new life east of the garden. But even in exile, God is not absent. Eve's words—"I have gotten a man with the help of the LORD"—reveal gratitude and perhaps even a glimmer of fulfilled prophecy. She may have hoped Cain was the promised "seed" from Genesis 3:15, the one who would crush the serpent's head. But as the story unfolds, that hope is painfully subverted.

The Fall continues to ripple through human history—not in abstract, but in blood. The consequences of sin do not remain within Eden's gates. They

92

follow mankind into the soil, the family, the city, and the culture. Yet even in judgment, we see divine mercy. Even in rebellion, God's design persists—twisted by sin, but not erased.

Genesis 4:1–5 – Worship, Sacrifice, and the Divide Between Brothers

"Now Abel was a keeper of sheep, and Cain a worker of the ground... And the LORD had regard for Abel and his offering, but for Cain and his offering He had no regard."

The contrast between Cain and Abel is not one of occupation—both shepherding and farming were noble and necessary in the ancient world. Rather, the difference lies in **attitude** and **intent**. Abel brings the **firstborn** of his flock—an act that signifies not just sacrifice, but trust. He gives the best. Cain, by contrast, brings only "some of the fruits"—an offering unmarked by faith or priority.

Hebrews 11:4 clarifies this distinction:

"By faith Abel offered to God a more acceptable sacrifice than Cain."

Abel's offering reflected faith and reverence; Cain's, indifference or presumption. This passage introduces a theme that will echo throughout Scripture: **God is not pleased with empty rituals**. He desires worship that springs from sincerity and truth.

"Those who worship Him must worship in spirit and truth."
— John 4:24

The contrast between the brothers is not about the external gift—it's about the internal posture. Worship, in biblical terms, is not about what we give, but **how and why** we give it. And when worship is divorced from humility, it becomes performance.

This moment also introduces the **possibility of divided humanity**: one line marked by faith and one by resistance. Cain and Abel foreshadow two paths—obedience and rebellion. The divide between them is not merely personal. It becomes **civilizational**.

Genesis 4:6–8 – Sin as a Predator and the First Murder

"Sin is crouching at the door. Its desire is for you, but you must rule over it."

Before Cain acts, God intervenes. He speaks directly—not in wrath, but in warning. This is the Bible's **first use of the word "sin"** (*chatta'ah* in Hebrew), and it's described not as an abstract idea, but as a **living force**—a predator lurking just outside the door of the soul. It waits. It desires. But it does not yet dominate. Cain still has a choice.

This portrayal resonates with **modern psychology and neuroscience**. Studies show that the human brain is wired for both impulse and restraint. The **prefrontal cortex**, responsible for reasoning and self-control, must actively regulate the emotional centers of the brain—particularly the **amygdala**, which drives anger, fear, and desire. Cain's situation mirrors this conflict. The desire is present— but so is the ability to resist.

Yet Cain refuses to master the impulse. The tension escalates, and instead of turning toward God, he turns against his brother. The text is devastating in its brevity:

"Cain rose up against his brother Abel and killed him."

No argument. No prelude. Only a premeditated act of violence. What began in envy ends in blood. The first death in human history is not a natural one—it is a murder. It is brother killing brother. And in that moment, the true depth of the Fall becomes evident: sin not only separates man from God—it separates man from man.

This is not merely an act of passion. It is the result of a heart already unrepentant. Cain refuses instruction. He resents correction. And in doing so, he becomes the first image of what will become **a pattern of rebellion** throughout the rest of Scripture.[18]

---

[18] Sapolsky, Robert M. Behave: The Biology of Humans at Our Best and Worst. New York: Penguin Press, 2017.

Genesis 4:9–16 – The Curse of the Wanderer and Divine Restraint

"Am I my brother's keeper?"

Cain's response to God is cold, deflective, even sarcastic.[19] He shows no sorrow. No repentance. Only evasion. In asking this question, Cain reveals how deeply sin has already taken root. He rejects responsibility. He severs the very notion of relational accountability. But the irony is heavy—because yes, he was meant to be his brother's keeper. That is the calling of humanity: to protect, preserve, and love one another. Cain's question is not philosophical—it is **a confession of his failure**.

God's response is firm, but merciful. He does not execute Cain, though justice would allow it. Instead, He pronounces a curse—not of death, but of **restlessness**. Cain becomes a wanderer. He is severed from the

---

[19] Bloom, Paul. Against Empathy: The Case for Rational Compassion. New York: Ecco, 2016.

ground, his source of identity and provision. He becomes unrooted, dislocated, **unsettled**.

This restlessness mirrors what psychologists observe in those who carry unconfessed guilt: emotional numbness, isolation, anxiety, and lack of rootedness. The burden of unresolved sin does not merely affect behavior—it affects identity.[20]

Yet even here, God places a **mark** on Cain—not as punishment, but as protection. This is an act of **divine restraint**. God allows Cain to live, though he deserves death. He ensures that vengeance will not escalate. In a world spiraling toward violence, this is God's first act of containing it.

Cain's exile becomes a metaphor for humanity's condition: **estranged, but not abandoned**.

---

[20] LeDoux, Joseph. Synaptic Self: How Our Brains Become Who We Are. New York: Viking, 2002

Genesis 4:17–24 – Cities, Culture, and the Cainite Legacy

> *"Cain knew his wife, and she conceived and bore Enoch. When he built a city, he called the name of the city after the name of his son…"*

Despite his curse, Cain continues to build. He marries, fathers children, and establishes a city. This is a remarkable contrast: the first builder of human civilization is also the first murderer. The legacy of Cain is not simplicity—it is complexity. It is the paradox of progress without peace.

The naming of the city after his son, Enoch, reflects a desire for permanence, legacy, and perhaps even **self-redemption**. Though exiled from Eden, Cain seeks to create his own ordered world—a man-made Eden. But his foundation is not righteousness; it is rebellion.

The genealogy that follows tells a fascinating story. In just a few generations, we see the rise of **culture**:

- **Jabal** is the father of livestock and nomadic herding—creating new systems of survival and mobility.
- **Jubal** invents musical instruments—string and wind—birthing **art and expression**.
- **Tubal-Cain** pioneers metallurgy—advancing **technology and craftsmanship**.

This brief but dense section of Genesis shows that even in exile, **human creativity flourishes**. The image of God in man—though marred—is not erased. The ability to make, mold, design, and innovate remains. Music, technology, and society are not foreign to the biblical worldview—they are part of God's common grace. Even in fallen hands, **creativity is a remnant of the divine design**.

Yet this same line also illustrates the **dark side of progress**. With innovation comes corruption. Lamech, a descendant of Cain, introduces **polygamy**—departing from God's design for marriage

(Genesis 2:24). And worse, he boasts of violence:

"I have killed a man for wounding me, a young man for striking me."

This is not justice—it is **vengeance**. Lamech's declaration reveals a worldview where power is exalted, and retaliation is celebrated. His twisted logic suggests that if Cain was avenged sevenfold, Lamech deserves seventy-sevenfold. The escalation is chilling. Violence is no longer hidden—it is **weaponized and institutionalized**.[21]

Thus, the line of Cain becomes a study in **civilization without righteousness**—a culture rich in achievement but devoid of moral restraint. The echo of Eden grows fainter.

Genesis 4:25–26 – Seth and the Hope of a New Line

"God has appointed for me another offspring instead of Abel, for Cain

---

[21] Pinker, Steven. The Better Angels of Our Nature: Why Violence Has Declined. New York: Viking, 2011.

killed him.”
— *Genesis 4:25*

Eve gives birth to another son—**Seth**, whose name means "appointed" or "granted." This is not merely a replacement in the emotional sense. Seth is a **restorative line**, a new branch in the human story through which the promise of redemption will continue.

From Seth's line will come Enosh, then Noah, then Abraham—and eventually, the Messiah Himself. What was lost in Abel is renewed in Seth. In this brief verse, we are reminded that **God's promise in Genesis 3:15 still stands**. The seed of the woman will not be extinguished.

> *"At that time people began to call upon the name of the LORD."*
> — *Genesis 4:26*

This closing verse is filled with quiet power. In contrast to the legacy of Cain—marked by pride, violence, and self-made identity—the line of Seth begins to return to **worship**. They "call upon the name of the LORD." This phrase signals more than prayer.

It marks the beginning of[22] **corporate worship**, of a community that remembers the Creator and seeks relationship with Him.[23]

This is the rise of a **faithful remnant**—a people within a corrupt world who do not forget the voice that once walked with them in Eden.

Conclusion: The Echoes of Eden in a Fractured World

Genesis 4 is a study in contrast. It shows two lines, two legacies, and two responses to the Fall. Cain represents alienation, self-reliance, and unrepentant pride. Seth represents restoration, dependence on God, and the long path to redemption.

In Cain, we see the origins of **secular civilization**—brilliant but broken. In Seth, we see the origins of **spiritual**

---

[22] Haidt, Jonathan. The Righteous Mind: Why Good People Are Divided by Politics and Religion. New York: Pantheon Books, 2012.

[23] Harari, Yuval Noah. Sapiens: A Brief History of Humankind. New York: Harper, 2015.

inheritance—humble but enduring. One builds cities. The other builds altars.

And yet, through it all, the **image of God persists**. Though sin spreads, so does creativity. Though the ground is cursed, life still rises from it. Music is played. Tools are forged. Livestock are tended. In both lines, we see the **power of design**—inherited from the Creator, even when misused.

But the true hope lies not in culture, but in covenant. The promise of redemption will not be fulfilled through invention, but through **incarnation**. The seed will come—not through Cain, but through Seth. Not through pride, but through faith.

Even in exile, the story moves forward. Eden may be behind us, but the echo of its design, and the promise of its return, continues to shape history. The way back to the garden will not be built with bricks—but with blood.

The cross is coming. And the seed is alive.

Chapter 6: Genesis 5 – Genealogy, Longevity, and the Line of Promise

> *"This is the book of the generations of Adam. When God created man, He made him in the likeness of God. Male and female He created them, and He blessed them and named them Man when they were created."*
> — *Genesis 5:1–2*

At first glance, Genesis 5 reads like a straightforward genealogy—a lineage from Adam to Noah. But beneath its surface lies a powerful reaffirmation of divine purpose and human identity in a world now marked by sin.

The chapter opens not merely with historical record but with theological reminder. Even after the Fall, even after exile and the first murder, **humanity is still made in the likeness of God**. The divine image, though tarnished by rebellion, is not erased. The echo of Eden resounds through every generation. God's blessing remains in effect. His name for mankind—*adam*—still affirms their identity and worth.

This introduction reframes the chapter that follows. What appears to be a repetitive list of lifespans and deaths is in fact a record of **preservation**. The lineage from Adam to Noah is not accidental—it is carefully sustained, safeguarded, and significant. It is the **line of promise**, through which the seed of the woman will ultimately come.

Genesis 5:3–32 – A Lineage of Life, and a Pattern of Death

The structure of Genesis 5 is rhythmic, almost liturgical:

"Adam lived... and he died."
"Seth lived... and he died."
"Enosh lived... and he died."

The phrase "and he died" becomes a refrain—a drumbeat reminding us of the **consequence of the Fall**. In Genesis 3, God warned that disobedience would lead to death. In Genesis 4, we saw the first murder. Now, in Genesis 5, we see that **death is no longer an exception—it is the rule**.

And yet, this genealogy is not a lament. It is a declaration that **God's design continues**. Despite mortality, the human story moves forward. Each name, each birth, each generational handoff is part of a divine thread. Sin does not stop the plan. Death does not erase the promise. Through quiet faithfulness, the **line of redemption marches on**.

The Mystery of Long Life Spans

One of the most remarkable features of Genesis 5 is the extreme longevity of its patriarchs. Adam lives 930 years, Seth 912, and Methuselah reaches 969 years—the longest human lifespan ever recorded in Scripture. For many, these ages seem implausible by modern standards, but several theories offer frameworks to understand them—both theologically and scientifically.

1. Textual Realism – Accepting the Ages at Face Value

This view holds that early human lifespans were literally as recorded, due to unique conditions in the pre-Flood world. Supporters of this view argue that environmental, genetic, and physiological factors may have been vastly different during Earth's earliest centuries. Genesis 1–6 implies that the world before the Flood was not only less corrupted morally, but **less degraded biologically and ecologically**.

2. The Hyperbaric Atmosphere Theory

One compelling scientific hypothesis involves the concept of a **hyperbaric environment**—a world with higher atmospheric pressure and increased oxygen concentration compared to today.

Creation scientists and researchers have proposed that Earth's early atmosphere may have functioned like a **global hyperbaric chamber**. This theory suggests a canopy or water-vapor barrier (referenced in Genesis

1:7, "waters above the firmament") that created:

- Increased **barometric pressure**, enhancing oxygen diffusion in human and animal tissues.
- Greater protection from **cosmic and ultraviolet radiation**, reducing cellular mutation and DNA damage.
- Stabilized, tropical-like conditions across the globe, fostering longevity and vitality.

Dr. Carl Baugh, founder of the Creation Evidence Museum, has been a notable proponent of this theory. His work with simulated hyperbaric chambers has demonstrated that **increased atmospheric pressure and oxygen saturation accelerate healing and boost physiological performance** in both animals and humans.

3. Scientific Support: Modern Hyperbaric Oxygen Therapy (HBOT)

Today, **hyperbaric oxygen therapy (HBOT)** is a medically recognized treatment used in hospitals and clinics worldwide. HBOT involves breathing 100% oxygen at pressures greater than atmospheric pressure, and has been scientifically shown to:

- **Stimulate stem cell release** and tissue regeneration (Thom et al., *Am J Physiol Heart Circ Physiol*, 2006)
- **Enhance immune function**, particularly white blood cell activity and pathogen elimination [24]
- **Reduce inflammation** and oxidative stress—key factors in aging [25]
- **Extend telomere length**, the caps of DNA linked to aging

---

[24] Heyboer, Marcus, et al. "The Use of Hyperbaric Oxygen Therapy in the Treatment of Radiation-Induced Soft Tissue Injury: A Review of the Literature." Undersea & Hyperbaric Medical Society Journal 44, no. 3 (2017): 191–200.
[25] Zelickson, Brian D., et al. "Evaluation of the effect of hyperbaric oxygen therapy on collagen synthesis and antioxidant levels in human skin in vivo." Journal of Investigative Dermatology Symposium Proceedings 13, no. 1 (2008): 35–39.

and cell death (Hadanny et al., *Aging*, 2020)

"We have demonstrated for the first time in humans that HBOT can **increase telomere length by over 20%** and decrease senescent cells by up to 37%."[26]

Telomeres shorten with age and stress, contributing to disease and mortality. Reversing this process, as seen in HBOT studies, provides **direct biological evidence** that environments high in oxygen and pressure can significantly slow or even **reverse markers of aging**.

If such results can be achieved through limited, controlled hyperbaric treatments, it stands to reason that **a naturally hyperbaric world**—with continuous atmospheric conditions of elevated oxygen and pressure—could produce **vastly longer lifespans**,

---

[26] Efrati, Shai, Amir Hadanny, Yair Fishlev Gabbay, and Yafit Berkovitz. "Hyperbaric oxygen therapy increases telomere length and decreases immunosenescence in isolated blood cells: A prospective trial." Aging 13, no. 15 (2020): 20935–20952.

faster healing, and greater biological resilience.

4. Genetic Entropy and Post-Flood Decline

Creation geneticist Dr. John Sanford, author of *Genetic Entropy: The Mystery of the Genome*, argues that **post-Fall genetic deterioration** plays a major role in declining human longevity. According to his research, mutations accumulate with each generation, leading to an irreversible loss of genomic integrity.

Sanford writes:

"The human genome is degenerating due to the relentless accumulation of mutations... This is consistent with the biblical record of **declining lifespans after the Flood**, as recorded in Genesis 11."[27]

Combining this with the hyperbaric theory suggests that **longevity**

---

[27] Sanford, John C. Genetic Entropy and the Mystery of the Genome. Lima, NY: FMS Publications, 2005.

declined after the Flood for two major reasons:

1. **Environmental collapse—** the destruction of the atmospheric canopy, resulting in greater radiation exposure, climate variability, and physiological stress.
2. **Genetic degradation—** mutations increasing over time, reducing cellular repair mechanisms and increasing disease susceptibility.

Conclusion

Whether one views the Genesis lifespans as literal, symbolic, or a mix of both, there is mounting scientific and textual evidence that **early humanity lived in a vastly different biological environment**. A hyperbaric world—richer in oxygen, shielded from radiation, and genetically closer to original perfection—could plausibly support the longevity described in Genesis 5.

Rather than dismissing these ages as myth, we are invited to consider them as **historical reflections of a world still echoing Eden**—a world where life endured longer, death came slower, and the effects of the Fall had not yet fully matured.

The Enoch Exception – A Life that Escaped Death

> *"Enoch walked with God, and he was not, for God took him."*
> — *Genesis 5:24*

Amid the steady drumbeat of birth and death, one name breaks the pattern: **Enoch**.

He lived 365 years—a relatively short life by Genesis 5 standards. But he did not die. Instead, we are told he "walked with God" and "was not, for God took him." The phrase "walked with God" occurs only a few times in Scripture. It implies **deep fellowship**, consistency, and closeness. Enoch's life was marked not just by longevity, but by intimacy with the Creator.

Hebrews 11:5 confirms this remarkable event:

"By faith Enoch was taken up so that he should not see death, and he was not found, because God had taken him. Now before he was taken he was commended as having pleased God."

Enoch becomes a **prototype of resurrection hope**. He is a preview of things to come:

- Like **Elijah,** who would also be taken up without tasting death (2 Kings 2:11).
- Like **Jesus,** who would conquer death itself and ascend to glory.
- Like the **Church,** which awaits the return of Christ, when "we who are alive… will be caught up… to meet the Lord in the air" (1 Thessalonians 4:17).

Enoch's disappearance is not just a mystery—it is a **message**: death is not the final word. There is another way. **Fellowship with God can transcend the grave**.

Methuselah and the Year of the Flood

"Methuselah lived 969 years, and he died."

Methuselah holds the record for human longevity. His name, possibly meaning "when he dies, it will be sent," has intrigued scholars for centuries. According to traditional biblical chronology, Methuselah died **in the same year the Flood came** (Genesis 7:11).

If this interpretation is correct, Methuselah's life becomes a **symbol of divine patience**. His 969 years are not merely a statistic—they are a measure of God's long-suffering. For nearly a millennium, judgment was withheld. His death marked the end of an era, and the beginning of a cataclysm.

This is consistent with the character of God described in 2 Peter 3:9:

"The Lord is not slow to fulfill His promise… but is patient toward you, not wishing that any should perish, but that all should reach repentance."

Methuselah becomes a **living delay of judgment**, a countdown clock hidden in the text—a reminder that even wrath is restrained by mercy.

Lamech, Noah, and the Prophetic Line

> *"Out of the ground that the LORD has cursed, this one shall bring us relief…"*
> — *Genesis 5:29*

When Lamech names his son Noah, it is not a casual decision. He gives him a name with purpose—**nacham**, meaning rest, comfort, or relief. In a world under curse, Lamech sees in Noah the hope of reversal.

And indeed, Noah will fulfill that prophecy—though not as Lamech may have imagined. Noah will enter a world of judgment and emerge into **a world renewed**. He will become a type of Christ:

- He will **enter the judgment waters** on behalf of a remnant.
- He will **preserve life** through obedience and faith.

- He will emerge into a new creation—one where **God's covenant is renewed**.

Noah's story, which begins in Genesis 5, becomes the turning point of early human history. His life is the bridge between **antediluvian corruption** and **post-Flood covenant**. He stands at the threshold of God's reset—an echo of Adam and a shadow of Christ.

Conclusion: Design Through the Generations

Genesis 5 may seem, at first glance, like a dull record of names and numbers. But it is anything but. It is a **genealogical witness** to God's continuing faithfulness. In the face of death, life continues. In the face of sin, God's image endures. In the face of rebellion, a remnant walks with God.

Enoch shows us that intimacy with the Creator is still possible. Methuselah reminds us of divine patience. Noah points forward to redemption through obedience. And the whole genealogy points to a future when **death will no longer have dominion**.

This is not a random list of ancient lifespans—it is a record of **design, preservation, and promise**. From Adam to Noah, each generation carried the seed of salvation. And that seed, passed through centuries, would one day become **flesh and dwell among us**.

Genesis 5 reminds us that history is not aimless. It is a thread of grace woven through time—line by line, name by name, moving ever forward toward redemption.

And it continues still—by design.

Chapter 7: Genesis 6–9 – The Flood, the Ark, and the Covenant of Renewal

> *"The LORD saw that the wickedness of man was great in the earth... and the LORD regretted that He had made man on the earth... But Noah found favor in the eyes of the LORD."*
> — *Genesis 6:5–8*

Genesis 6:1–8 – Corruption, Violence, and Divine Grief

Humanity, which began in a garden of order and innocence, now saturates the earth with **violence, corruption, and moral collapse**. The divine observation in verse 5 is staggering in scope:

"Every intention of the thoughts of his heart was only evil continually." This is not just an indictment of actions—it's a diagnosis of the inner life. Humanity's imagination, desire, and moral compass are fundamentally distorted.

God's response is not detached. He does not observe with cold judgment but with sorrow:

"The LORD regretted that He had made man…"
This regret is not divine error—it is divine grief. It is the pain of a Creator watching His image-bearers unravel.

Yet even here, a thread of grace emerges:

"But Noah found favor in the eyes of the LORD."
This is the **first explicit mention of grace** in the Bible. Noah is not perfect, but he is faithful—blameless in his generation. Like Enoch before him, Noah **walked with God**, a rare and intimate phrase that denotes fellowship, trust, and obedience. Amid global wickedness, one man shines as a remnant of righteousness.

Genesis 6:9–22 – The Ark: A Vessel of Judgment and Salvation

God gives Noah instructions for an **ark**—not a vessel of speed or elegance, but of endurance and preservation. The specifications are detailed:

- **300 cubits** long (~450 ft)

- **50 cubits** wide (~75 ft)

- **30 cubits** high (~45 ft)

These proportions correspond remarkably with modern maritime engineering. Naval architects such as **Tim Lovett** have shown that these dimensions provide optimal **stability, buoyancy, and structural strength**, especially in turbulent conditions. In fact, modern cargo barges often follow similar ratios—designed not to navigate swiftly, but to survive storms.[28]

Scientific Considerations:

- The ark's internal volume would exceed **1.5 million cubic feet**, roughly the capacity of **500 railroad boxcars**—enough to house thousands of animals, particularly when considering juvenile representatives and hibernating species.

- The use of **gopher wood** and "pitch" (a resinous waterproofing material) points to **real-world construction methods** suitable for long-term flotation.

Why a Global Flood?

Geological evidence points to the plausibility of a large-scale flood event:

---

[28] Lovett, Tim. Noah's Ark: A Feasibility Study. Green Forest, AR: Master Books, 2008.

- **Sedimentary rock layers** across continents indicate rapid deposition of water-borne material.[29]

- **Marine fossils** have been found on mountaintops around the world—including the Himalayas.[30]

- **Mass extinction layers** and fossil graveyards suggest catastrophic die-offs involving water.

While secular scientists may argue for localized flooding, the **ubiquity of global flood legends** (over 200 independent accounts from cultures worldwide) suggests a deeply embedded **collective memory**. These

[29] Austin, Steven A., et al. "Catastrophic Plate Tectonics: A Global Flood Model of Earth History." In Proceedings of the Third International Conference on Creationism, 609–621. Pittsburgh: Creation Science Fellowship, 1994.
[30] Snelling, Andrew A. Earth's Catastrophic Past: Geology, Creation & the Flood, Vol. 1. Dallas: Institute for Creation Research, 2009

stories, from Mesopotamia to Mesoamerica, often contain striking parallels: divine judgment, a chosen survivor, a great vessel, and a new beginning.[31]

Genesis offers not merely a story, but **a historical anchor**—God's moral response to unchecked wickedness, and His provision of salvation through obedience.

Genesis 7 – The Waters Rise

"All the fountains of the great deep burst forth, and the windows of the heavens were opened."
This verse describes a dual-source cataclysm:

- **Fountains of the deep** likely refer to seismic and volcanic eruptions, possibly linked to tectonic activity and the

---

[31] Fritz, Paul N. Flood Legends: Global Clues of a Common Event. Green Forest, AR: Master Books, 2009.

sudden release of subterranean water (see hydroplate theory).

- **Windows of heaven** represent sustained, torrential rainfall, perhaps for the first time in Earth's history.

Rain falls for **40 days**, but the water **covers the earth for over 150 days**. This wasn't a brief storm—it was an **extended geological reset**, reshaping the planet's surface. The ark rises, untouched by destruction below, carried solely by God's provision.

Genesis 8 – The Waters Recede and the Altar is Raised

"God remembered Noah… and made a wind blow over the earth."
In Hebrew, to "remember" (זָכַר) is not about recollection—it is **covenantal action**. God moves on behalf of Noah because of His promise, not because He had forgotten.

As the waters recede, the ark **comes to rest on the mountains of Ararat**—a region straddling modern-day **Turkey and Armenia**. Mount Ararat has long been considered a prime candidate for the ark's resting place, with numerous expeditions, ancient testimonies, and enduring tradition pointing to this mountainous zone.

Noah's first act upon leaving the ark is not survival—it is **worship**. He builds an **altar**, offering a burnt sacrifice. God responds with favor, declaring:

"I will never again curse the ground because of man…"

The flood ends with **sacrifice**, not celebration. The pattern is clear: **judgment yields to mercy, and mercy flows from worship**.

Genesis 9 – Covenant, Rainbow, and the Renewal of Design

"Behold, I establish My covenant with you and your offspring after you… never again shall there be a flood to destroy the earth."

With this divine statement, God initiates the **Noahic Covenant**, a binding promise not just to Noah, but to **all creation**. The **sign of the covenant** is the **rainbow**—a visible reminder of invisible grace.

New guidelines are established:

- **Human life is sacred**—the shedding of blood demands justice.

- **Meat is permitted**, but blood remains prohibited, preserving the sanctity of life.

- The original mandate is reaffirmed: "Be fruitful and multiply."

The rainbow is not a meteorological curiosity—it is **a cosmic declaration**.

129

It marks the end of judgment and the beginning of renewed design. It tells a broken world: there is **hope beyond the flood**.

Scientific Echoes of the Flood

Creation scientists have identified several lines of physical evidence consistent with a global flood:

- **Sedimentary layers** spanning continents, often with no signs of slow, uniform accumulation.

- **Polystrate fossils**—trees that extend upright through multiple rock layers—impossible to explain by gradual burial.[32]

- **Fossil graveyards** with mixed marine and terrestrial life, suggesting violent, watery upheaval.

---

[32] Morris, John D. The Young Earth: The Real History of the Earth – Past, Present, and Future. Green Forest, AR: Master Books, 2007.

- **Genetic bottlenecks** in human and animal DNA, hinting at recent, severe population reductions consistent with the biblical flood timeline.[33]

Experts such as **Dr. Andrew Snelling** (Answers in Genesis) and **Dr. Kurt Wise** argue that many geological features—such as the **Grand Canyon**, coal beds, and large-scale erosion patterns—are better explained by **catastrophic flood geology** than by uniformitarian models.

The Rainbow: Was It Visible Before the Flood?

"I have set my bow in the cloud…"
— Genesis 9:13

This verse introduces one of Scripture's most debated questions: Was the rainbow newly created after

---

[33] Jeanson, Nathaniel T. Traced: Human DNA's Big Surprise. Green Forest, AR: Master Books, 2022.

the flood, or simply given new meaning?

To form a rainbow, three elements are required:

1. Direct **sunlight**

2. **Water droplets** in the atmosphere

3. An **observer** positioned with their back to the sun.

Rainbows result from light refracting and reflecting within water droplets, producing a circular spectrum. But if conditions for rain or sunlight were different, rainbows may not have occurred—or been visible.

Pre-Flood Atmospheric Conditions:

Creationist models propose that the pre-Flood world had:

1. A **water vapor canopy**, filtering UV light and diffusing sunlight.

2. A **mist-based hydrological cycle** (Genesis 2:5–6) with **no rain**.

3. **Uniform temperatures and high atmospheric pressure**, which would reduce the contrast necessary for optical phenomena like rainbows.

If these models are accurate, then the post-Flood environment—with exposed sunlight, variable weather, and rainfall—was **the first time conditions allowed rainbows to form**.

Moreover, the Hebrew phrase "I have set" (נָתַתִּי) suggests placement, not just designation. Even if the rainbow could theoretically exist before, it now **functions as a divine sign**, much like the bread and wine in the Lord's Supper. **A natural element becomes a spiritual symbol**.

Scientific Support:

- Studies in **cloud physics** and **light scattering** confirm that dense atmospheric layers can obscure optical clarity.[34]

- Volcanic eruptions and atmospheric dust provide modern analogs for the **opacity** that may have existed pre-Flood.

- After a cataclysm, the sky would be clearer, more open to direct light—**ideal for rainbow visibility**.

Conclusion: A Bow in the Sky, A Covenant on the Earth

The Flood is not merely a tale of destruction—it is a revelation of divine justice and divine mercy. In Noah, we see a man of obedience, whose faith becomes a vessel for salvation. In the ark, we see a shadow of Christ—**a single door, a sealed**

---

[34] Zhang, Pengfei, and Anthony R. Lupo. "The Influence of Water Vapor and Clouds on Earth's Energy Budget." Atmosphere 10, no. 4 (2019): 181.

refuge, a place of shelter while judgment rages outside.

And in the rainbow, we see the aftermath: not just the end of a storm, but the **beginning of a new promise**. A **bow of war**, now unstrung and pointed away from the earth, hangs in the heavens—a visible assurance that **mercy triumphs over judgment**.

The world was judged. The world was washed. But from the water rose worship, covenant, and the enduring design of God. From Noah's altar to Christ's cross, the pattern is clear: **judgment yields to sacrifice, and sacrifice restores design**.

And still today, when storms pass and light breaks through, the bow appears. And it speaks.

Chapter 8: Genesis 10–11 – Nations, Babel, and the Scattered Image of God

> *"These are the generations of the sons of Noah... from these the nations spread abroad on the earth after the flood."*
> — *Genesis 10:1, 32*

Genesis 10 – The Table of Nations: The Blueprint of Civilization

Genesis 10 is not a myth. It is a map. Known as the "Table of Nations," this chapter is one of the oldest and most comprehensive ethnographic records in the world. It catalogs seventy lineages descending from Noah's sons—**Shem**, **Ham**, and **Japheth**—who became the fathers of nations across the ancient world. These were not merely family trees; they were **foundations of**

civilization, establishing the identities of real people in real regions.

In contrast to the self-exalting genealogies of Mesopotamian kings or Egyptian pharaohs, the Table of Nations is:

- **Historically grounded**: Many names correspond to tribes, peoples, or regions attested in archaeology and ancient texts.
- **Structured by geography and ethnicity**, not by imperial power or mythological embellishment.
- **Theologically inclusive**: It does not center Israel; rather, it presents **all peoples as equally descended from Noah**, reflecting a unified human origin.

Scientific and Historical Alignment

The Table's accuracy is remarkable. The descendants of **Japheth** populated the Indo-European sphere—**Gomer** correlates with the **Cimmerians and Gauls**, **Magog** with the **Scythians**, **Madai** with the

Medes, and **Javan** with the **Greeks** (Ionians). These names appear in ancient inscriptions from Assyria, Greece, and Persia.

Ham's line settled across Africa and parts of the Near East. **Cush** represents **Nubians and Ethiopians**, **Mizraim** is the Hebrew name for **Egypt**, **Put** correlates with **Libya**, and **Canaan** with the Levantine peoples.

Shem's lineage is the most theologically emphasized, as it leads to **Abraham**. His sons include **Elam** (Persians), **Asshur** (Assyrians), **Aram** (Arameans/Syrians), and **Eber**, from whom the term "Hebrew" derives.

Modern Genetic Confirmation

Genetics affirms the biblical claim of a **shared human ancestry**. Studies in **mitochondrial DNA (mtDNA)** and **Y-chromosome tracing** confirm that all living humans descend from a **single woman (mt-Eve)** and a **single man (Y-Adam)** who lived in the relatively recent past—most likely centered around the **Middle East or northeastern Africa**, exactly where

Genesis locates the origin post-Flood.[35]

Creation geneticist **Dr. Nathaniel Jeanson**, in his book *Traced: Human DNA's Big Surprise* (2022), demonstrates how global patterns of **Y-chromosome variation** reveal a rapid population expansion originating in the **biblical time frame**. Contrary to evolutionary assumptions of deep, gradual divergence over hundreds of thousands of years, Jeanson's research suggests **a rapid, post-Babel dispersal**, aligning with Genesis 10–11.[36]

Even secular population geneticists note that human genetic diversity shows a **"bottleneck effect"**—a severe reduction followed by expansion—exactly what we would expect after the Flood.

---

[35] Cann, Rebecca L., Mark Stoneking, and Allan C. Wilson. "Mitochondrial DNA and Human Evolution." Nature 325 (1987): 31–36.

[36] Jeanson, Nathaniel T. Traced: Human DNA's Big Surprise. Green Forest, AR: Master Books, 2022.

Genesis 11:1–4 – The Tower of Babel: Unity Without God

> *"Now the whole earth had one language and the same words... Then they said, 'Come, let us build ourselves a city and a tower with its top in the heavens... and make a name for ourselves."*

While Genesis 10 zooms out to show the spread of people, Genesis 11 zooms in to show the spiritual heart of the problem. **The people are united—but not in righteousness.** Their unity becomes rebellion.

Instead of spreading across the earth, as God commanded in Genesis 9:1 ("Be fruitful and multiply and fill the earth"), they choose to **centralize**, **settle**, and **control**. Their goal is explicit:

"Let us make a name for ourselves."

This is not just construction, it is defiance. The tower is not a stairway to God—it is an **attempt to bring God down** or **domesticate the divine**. Most scholars believe the tower was a **ziqqurat**—a stepped pyramid common in ancient Mesopotamia, symbolizing human attempts to access divine power on their own terms.

Just as Adam and Eve grasped at divine knowledge in Eden, the builders of Babel grasp at **divine stature**, not through obedience, but through **ambition**.

The project is a kind of **proto-globalism**—not technological or economic, but spiritual. It is the dream of **humanity exalted without humility**, a vision of the future with **God replaced by man**.

Genesis 11:5–9 – The Scattering and the Birth of Nations

> *"And the LORD came down to see the city and the tower... and said, 'Come, let us go down and confuse their language...'"*

The irony is deliberate. Humanity attempts to build up to heaven, and yet God must "come down" to even regard their work. His descent is not out of ignorance—it is **judicial involvement**.

God's response is swift and surgical. He confuses their language, not to punish arbitrarily, but to **protect humanity from its own pride**. Unified rebellion would have produced centralized tyranny—a single power unchecked by difference. By introducing linguistic diversity, God forces **dispersion**—ensuring that no single culture dominates all others.

This scattering is not the end of civilization—it is the **birth of true nations**. From many tongues and tribes, cultures begin to bloom. Language becomes not only a means of communication, but a **container for identity, worldview, and memory**.

Scientific Corroboration

- **Linguistics**: Modern languages derive from a small number of **proto-language families** (e.g., Afro-Asiatic, Indo-European, Sino-Tibetan), suggesting a common root consistent with a single original language fragmented in time.[37]

- **Genetics**: Human genetic variation fans out in patterns from the **Near East**, not Africa as commonly taught, supporting the **biblical center of dispersion**.
- **Cultural Memory**: Mesopotamian myths such as **Enmerkar and the Lord of Aratta** reference a time when all people shared one tongue until the gods intervened. Indian, Mayan, and Polynesian traditions also include stories of ancient unity disrupted by divine action.

Linguist **Dr. Stephen Taylor**, writing for the Linguistic Society of America, observes that **forced language fragmentation**—especially following political or spiritual upheaval—is a well-documented phenomenon. When large populations split suddenly, **dialects diverge rapidly**, becoming

---

[37] Frazer, James G. The Golden Bough: A Study in Comparative Religion. New York: Macmillan, 1922.

distinct tongues within a few generations.

God's Sovereign Design Through Scattering

Though the scattering of Babel appears chaotic, it is **a redemptive act**. Like the curse in Eden, it restrains evil while preparing for grace.

Through dispersion, God:

- **Prevents totalitarianism**, no single power can dominate the earth.
- **Fosters cultural resilience**— diverse groups adapt to different lands, forming new societies.
- **Sets the stage for redemption**—from one man (Abraham, in the next chapter), God will call a nation through whom all nations will be blessed.

God uses the very thing He divided— **language**—as a tool for reunification at Pentecost (Acts 2), when the Spirit enables people to hear the gospel in their native tongue. And ultimately,

Revelation 7:9 foresees the day when every language and tribe stands united—not in rebellion, but in worship.

The confusion of Babel is not the end of the story—it is a **pivot point in the plan of redemption**.

Conclusion: The Story Behind the Map

Geneses 10 and 11 are not relics of ancient history—they are blueprints of **who we are** and **how we got here**. The Table of Nations affirms a common origin and divine oversight into the spread of humanity. The story of Babel warns against centralized pride and the illusion of self-made unity.

But through it all, **God's design persists**. He formed every nation, appointed its times and boundaries (Acts 17:26), and embedded His image in every language, culture, and race. What sin scatters, grace gathers.

The tower of Babel fell. But the kingdom of God will not.
The languages split. But the Word

became flesh.
And even from brokenness, God
continues to build—by design.

Chapter 9: How Did Moses Know? – Divine Revelation and the Scientific Precision of Genesis

"The secret things belong to the LORD our God, but the things that are revealed belong to us and to our children forever..."
— *Deuteronomy 29:29*

Introduction: The Impossible Insight

How did Moses—a man born in the ancient Near East, trained in Egyptian wisdom but with no modern instruments—record an origin narrative that aligns with scientific discoveries made only in the last few centuries? How could he, living in the 15th century BC, have described a structured creation, the formation of life, the dispersion of nations, and the emergence of languages with such uncanny accuracy?

The book of Genesis, particularly chapters 1 through 11, contains profound truths that are increasingly affirmed by modern disciplines: cosmology, biology, geology, and linguistics. These are not the random myths of primitive people. They are structured, reasoned, and astonishingly aligned with reality. And that raises the central question: **How did Moses know?**

The answer is embedded in the heart of Scripture. Moses did not write Genesis as a product of guesswork, cultural osmosis, or intellectual synthesis. He wrote under divine inspiration—a prophet, not a philosopher, a revealer, not a speculator. The precision of Genesis points not to the ingenuity of man, but to the **intervention of God**.

"How Did Moses Know?"

An Extended Narrative Exploration of Revelation, Science, and the Origins of Genesis

Part I: The Paradox of Precision

The man stood alone at the base of the mountain, still veiled by the smoke of Sinai's glory. Behind him, the camp of Israel murmured, waiting for words from the one who had entered the cloud and lived. Before him lay not only a nation—but a legacy. He had been given the task to teach slaves how to live as sons, to order a society around the holy, and to pen the story of beginnings. But how?

Moses, the adopted son of Pharaoh's daughter, was educated in the courts of Egypt—familiar with the cosmologies of Ra and Atum, the myths of watery chaos and gods of the sun. Yet what he would write in Genesis bore no resemblance to the confused pantheon of Egypt. The Hebrew Scriptures did not present gods emerging from nature—they presented nature as the product of the singular, eternal God.

"In the beginning God created the heaven and the earth…"

With that single sentence, Moses declared a truth so fundamental, so revolutionary, that it would take humanity thousands of years to catch up. He declared that time, space, and matter had a beginning. That creation was not cyclical, as the Babylonians claimed, nor eternal, as the Greeks thought, but initiated—spoke into being by One outside of it all.

But how did Moses know?

Part II: Beyond the Reach of Man

In modern times, with radio telescopes scanning the skies and physicists exploring quantum fields, the evidence for the beginning of the universe has become unavoidable. The Big Bang theory, though secular in articulation, affirms the idea that space, time, and matter had an origin—a moment when all that is sprang into being from nothing. This singularity, as cosmologists call it, is not just a concept. It is a measured reality, traced in the echo of cosmic background radiation and the expansion of galaxies.[38]

Yet Moses had no telescope. He had no laboratory. He stood in the sands of the Sinai, 3,500 years removed from modern science. And he wrote of a beginning.

He described light coming before the sun. He described the waters gathered, the land rising, the plants appearing before animals. He spoke of mankind as formed from the dust of the ground—an image now remarkably echoed by the periodic table, which tells us that our bodies are composed of elements forged in stars and found in soil.[39]

He described life as bearing seed "after its kind," and today, we know that genetic replication ensures that

---

[38] Penzias, Arno A., and Robert W. Wilson. "A Measurement of Excess Antenna Temperature at 4080 Mc/s." The Astrophysical Journal 142 (1965): 419–421.
[39] Sagan, Carl. Cosmos. New York: Random House, 1980.

species reproduce with remarkable fidelity within their bounds.[40]

He described man as made in the image of God—distinct from the animals, rational and moral, relational and creative. And every branch of psychology, anthropology, and philosophy that grapples with human uniqueness affirms that we are *not merely advanced animals.*

These are not the words of a tribal myth-maker. These are revelations ahead of their time.

So again we ask: How did Moses know?

Part III: The Limit of Human Inference

Skeptics argue that Moses must have borrowed—perhaps from Sumerian epics like Enuma Elish, or from Egyptian cosmology. But those ancient myths are filled with chaotic deities, violent origins, and incoherent cosmologies. Genesis is different. It is

---

[40] Sanford, John C. Genetic Entropy and the Mystery of the Genome. Lima, NY: FMS Publications, 2005

monotheistic, ordered, logical, and moral. There is no war among the gods. There is no divine genealogy. There is only *God*—one, sovereign, and separate from creation.

Others claim that Moses invented the story to shape Israel's identity. But if so, he chose the hardest path. He wrote of the failures of his ancestors, of humanity's fall, of divine judgment by flood, and of the confusion of languages. He did not exalt his people. He humbled them with truth.

No amount of observation or philosophy would have led Moses to write Genesis 1–11. Without instruments, he could not have deduced the expansion of the universe. Without geology, he could not have discerned the deep time of rock layers and the memory of a global flood. Without microbiology, he could not have envisioned DNA encoding information in every cell.

But the Genesis account anticipates these very realities—not with technical jargon, but with profound conceptual accuracy.

Why?

Because Moses was not inventing. He was revealing.

Part IV: The Prophet, Not the Philosopher

Scripture tells us exactly how Moses knew.

Numbers 12:6–8 declares that while God spoke to other prophets in visions and dreams, He spoke to Moses *face to face*, clearly, not in riddles. Moses was not merely a wise man or a cultural compiler. He was the chosen instrument of divine revelation.

The Hebrew word for "revelation" (*galah*) carries the sense of uncovering something hidden. Moses did not discover truth. It was disclosed to him.

On Mount Sinai, where the law was given, God also gave Moses the beginning. In some way beyond our understanding, Moses was shown what preceded him—not as myth, but as revelation. He saw what Adam could not record. He beheld what no man had witnessed.

Genesis is not a philosophical meditation on origins. It is a historical declaration. It is prophecy in narrative form.

And so, Genesis 1–11 reads not like the fables of Babylon or the sagas of Greece, but like a structured, intentional unveiling of reality—both physical and spiritual.

Part V: The Structure of Revelation

The precision of Genesis is not just in what it says, but in how it says it.

Day One: Light. Day Two: Separation of waters. Day Three: Land and vegetation. Day Four: Luminaries in the heavens. Day Five: Fish and birds. Day Six: Land animals and mankind.

The structure is poetic yet ordered, reflective yet informative. It mirrors the logic of forming and filling. The first three days form the environment; the next three days fill it with life.

This literary architecture echoes the architecture of creation itself. The cosmos is not random. It is lawful. It operates by constants—gravitational, electromagnetic, atomic. The universe is mathematically elegant, suggesting not just power, but design.

And Moses—without Newton, Einstein, or Hubble—wrote of this design.

He also captured truths now affirmed by other disciplines:

- **Linguistics** has shown that languages do not evolve from chaos but emerge fully formed and then devolve in complexity. Genesis 11 describes the dispersion of languages at Babel—a confounding act by God that aligns with the sudden appearance of distinct language families in human history.[41]
- **Geology** speaks of massive sedimentary layers, vast fossil beds, and marine deposits on mountaintops—all suggestive of cataclysmic flood conditions. Genesis 6–9

---

[41] Campbell, Lyle. Historical Linguistics: An Introduction. Cambridge: MIT Press, 2004.

records a global flood, not as myth, but as judgment.

- **Genetics** confirms a single human race, genetically unified and recently diverged. Genesis 1 and 3 present humanity as descending from a single couple—created, not evolved.
- **Anthropology** finds shared myths of creation and flood across cultures, echoing a shared memory. Genesis claims to be the source event.

Once again: How did Moses know?

Part VI: The Voice in the Cloud

It must have haunted him, in the best way. Not fearfully—but reverently.

The memory of the burning bush that did not consume. The trembling mountain. The voice that thundered yet filled him with peace. The stone tablets etched by divine hand. The hours he spent alone with God, returning with his face aglow. The tent of meeting, where he spoke with God "as a man speaks to his friend" (Exodus 33:11).

This was not hallucination. This was not myth-making. This was revelation.

And through that revelation, Moses not only led a people out of Egypt—he led the world to the truth about its origins.

When he wrote Genesis, Moses was not reflecting on ancient Near Eastern myths. He was not crafting an origin tale to rival other tribes. He was faithfully recording what had been shown to him by the Creator Himself.

That is why Genesis holds its power.

That is why, across millennia, its words remain unmatched.

Part VII: The Witness of Time

Today, with satellites orbiting above and genome sequencers humming in labs, we can see deeper into the cosmos and further into life than ever before. And what do we find?

We find fine-tuning in the constants of physics that makes life possible.[42]

We find a universe that had a beginning.

We find encoded information in DNA—language, logic, code.

We find boundaries between species, fossil evidence of sudden appearances, and geologic formations consistent with rapid, water-driven events.

We find in human behavior a moral compass, a longing for eternity, a capacity for worship.

And standing behind all of this is a single voice—speaking through Moses, across time:

"In the beginning, God created…"

---

[42] Rees, Martin. Just Six Numbers: The Deep Forces That Shape the Universe. New York: Basic Books, 2001.

Conclusion: From Revelation to Recognition

How did Moses know?

Because God told him.

Because divine revelation is not limited by the tools of man. Because the Author of creation is also the Author of truth. And in choosing Moses—a shepherd raised in palaces, humbled in deserts, and transformed by glory—God chose a vessel through whom both law and light would be given.

Genesis is not ancient guesswork. It is divine disclosure.

The more we uncover through science, the more we return to what Moses already said.

And so, the question is no longer just, *How did Moses know?*

The question is: *Are we willing to listen?*

The Context of Ancient Creation Myths

To understand how radical Genesis is, we must compare it to its cultural backdrop. Moses lived in a world teeming with creation stories—none of which resemble the orderly, monotheistic account found in Genesis. The ancient Near East was rich in mythology, but poor in coherence.

Babylonian Enuma Elish

In this tale, the god Marduk slays the goddess Tiamat, splitting her body to form the heavens and the earth. Creation is birthed from violence, chaos, and divine rivalry. Humanity is created from the blood of a rebel god—designed to be slaves to the pantheon.

Egyptian Cosmogonies

In Egyptian tradition, the sun god Ra emerges from a cosmic egg or a primordial lotus. He produces other deities through various bodily functions—some from spitting, others from tears or sweat. Creation is the byproduct of divine biology, not divine will.

Sumerian Texts

The world forms from a mingling of saltwater and freshwater gods. Humanity is created to relieve the gods of agricultural labor.

None of these accounts present a moral, transcendent God who speaks reality into existence. None describe a structured, sequential creation culminating in human dignity. **Genesis stands alone**—offering not myth, but meaning. Not chaos, but cosmos. Moses' writing is not derivative; it is **revolutionary**.

Cosmology and the Structure of the Universe

"In the beginning, God created the heavens and the earth."
— *Genesis 1:1*

This first verse of Scripture captures what cosmologists only recently confirmed through the **Big Bang theory**: that time, space, and matter had a single origin point. Genesis 1:1 is a remarkably concise summary of modern physics:

- "In the beginning" implies **time**.
- "The heavens" implies **space**.
- "The earth" implies **matter**.

From this opening statement, Genesis unfolds a **chronological sequence** of creation events that mirrors the broad strokes of cosmic development:

- **Light precedes stars** (Genesis 1:3)—just as photons preceded stellar formation in the early universe.
- **The atmosphere forms** (1:6–8), dividing waters above from below—aligning with the gradual cooling and layering of the Earth.
- **Dry land appears from the ocean** (1:9)—a process echoed in plate tectonics and the emergence of continents.

One particularly striking detail is found in Genesis 1:14–19, where the **sun, moon, and stars become visible** only after the establishment of land and vegetation. This matches the

Great Oxygenation Event in Earth history, when cyanobacteria began to flood the atmosphere with oxygen, dissipating the dense haze and making the sky transparent to celestial bodies.

Moses could not have known any of this from Egyptian science or Babylonian lore. The cosmological structure of Genesis is not only unique—it is **astoundingly accurate**.

Biological Sequence and Plant-Animal Life

Genesis 1:11–25 describes the emergence of life in **a logical, layered sequence**:

1. Plants
2. Marine creatures
3. Birds
4. Land animals
5. Humans

Modern science confirms that **photosynthetic life** had to exist before animal life. Oxygenation of the atmosphere was critical to the development of complex organisms. Without early plants—particularly cyanobacteria and algae—the Earth could not support respiration-based life.

Furthermore, Genesis consistently uses the phrase "**according to their kinds**," indicating biological boundaries. This aligns with modern observations of **genetic limits**: species reproduce within fixed categories, showing variation (microevolution) but not indefinite transformation (macroevolution). The fossil record and genetics reveal **stasis within forms**, not transitional chaos.

Genesis doesn't describe a random process. It presents **design, order, and replication**—the very foundations of life sciences today.

Human Origins and Uniqueness

"Let us make man in our image, after our likeness."
— *Genesis 1:26*

The creation of man is set apart in both form and purpose. Unlike animals, humans are endowed with **the image of God**—a theological truth confirmed by anthropological uniqueness.

Humans alone:

- Possess **abstract reasoning and symbolic language**.
- Create **art, music, mathematics, and metaphysics**.
- Establish **moral codes**, religions, and systems of justice.
- Seek **eternity**, purpose, and transcendence.

Genesis 2:7 adds scientific resonance:

"Then the LORD God formed the man of dust from the ground…" Chemically, the human body is composed of the same elements found in soil: carbon, hydrogen, nitrogen, calcium, iron. Carl Sagan admitted, "We are made of star-stuff." But Genesis said it first—and more meaningfully.

"…and breathed into his nostrils the breath of life."
Science can explain respiration, but

not **consciousness**. Neuroscience still struggles to define **the self**, and no theory can explain the origin of **free will, creativity, or the soul**. Genesis reveals that these are not emergent properties—they are divine gifts.

Noah's Flood and Geological Memory

Genesis 6–9 records a flood of global scope, divine purpose, and redemptive outcome. While critics often dismiss the flood as myth, geological evidence suggests otherwise:

- **Fossil layering across continents** is consistent with rapid burial in water.
- **Polystrate trees**, standing upright through multiple sediment layers, require rapid deposition.
- **Marine fossils atop mountains**, including the Himalayas, suggest massive water movement.
- Over **200 global flood legends**, from Aboriginal Australia to Native America,

preserve cultural memory of a catastrophic deluge.[43]

Genesis provides precise timing: 40 days of rain, 150 days of water coverage, and a resting place in the "mountains of Ararat." Even the **rainbow** as a covenant sign reflects atmospheric transformation. Genesis 2:5–6 describes a pre-Flood world without rain—suggesting a vapor canopy or mist-based hydrology. Only after the flood did the **conditions for rainbow formation**—sunlight, raindrops, and clear skies—come together.

Genesis explains the **why** behind what geology and meteorology have observed.

Genesis 10–11: Dispersion, Language, and Nations

Genesis 10 gives the **Table of Nations**—70 distinct family lines from Noah's sons. Genesis 11 tells the story of **Babel**, where linguistic

---

[43] Campbell, Lyle. Historical Linguistics: An Introduction. Cambridge: MIT Press, 2004.

confusion scattered people across the earth.

Science now affirms the structure and effects of this event:

- Linguists trace all modern tongues back to **proto-languages**, suggesting sudden divergence from a common root.
- Genetics reveals that human diversity **clusters by region**, originating from the **Middle East**.
- Archaeology confirms early urban centers and **ziqqurats** in Mesopotamia, consistent with Babel's description.

The Tower of Babel narrative is not a tale of punishment alone—it is the **origin story of cultural diversity**. And again, Moses records it with theological depth and anthropological accuracy.

Could Moses Have Guessed All This?

The Fingerprint of Revelation

The statistical odds of Moses independently deducing, without error, the scientific order of the universe's origin, the geological sequences of a global flood event, the progressive complexity of life, and the global dispersion of languages—*without access to modern instrumentation or empirical methodology*—approach absolute zero. It would be the equivalent of striking a perfect chord on a piano without having ever seen a keyboard, read sheet music, or heard a single note.

Moses possessed no telescope to peer into the heavens and witness cosmic expansion. He had no microscope to discern the intricacies of cellular life or the double-helix of DNA. He had no access to seismographs to map tectonic shifts or to satellite imaging to trace the migration of early civilizations. He could not analyze the genome or carbon date a fossil. He lacked even the linguistic tools of comparative philology, let alone the anthropological frameworks required

170

to trace the global divergence of languages following a single point of origin.

And yet, the first eleven chapters of Genesis record—with astonishing simplicity and elegance—truths that align with the core conclusions of astrophysics, biology, geology, and linguistics. Genesis speaks of a beginning to time and matter (Genesis 1:1), a concept that would not be verified until the 20th century with the formulation of the Big Bang theory. It describes light preceding the formation of the sun (Genesis 1:3–16), now understood in cosmology as a reality due to photon release and cosmic background radiation. It traces an ascending order of life—plants, aquatic life, birds, mammals, and finally man (Genesis 1:11–27)—a pattern echoed in the fossil record and biological complexity.

During an age when ancient myths explained the universe through pantheons of warring deities, serpents carved from chaos, and mountains birthed from divine corpses, Moses wrote a singular, unified account—a cosmos brought forth not by conflict, but by command. Genesis does not

171

resemble the mythologies of Egypt, Mesopotamia, or Canaan. In the Egyptian cosmogony, the sun god Ra is born from a lotus flower rising out of primordial waters. In Babylonian tales, Marduk carves the heavens and earth from the slain body of Tiamat. These are cyclical, chaotic stories. In contrast, Genesis is linear, ordered, declarative. *"And God said... and it was so."*

Where other narratives offered gods of sun and soil—elemental deities bound to creation—Moses unveiled a transcendent Creator: the **Author** of time, the **Initiator** of space, the **Sustainer** of matter, and the **Giver** of life. No longer was man to worship the creation; he was to revere the One who stood outside of it and called it into being by the power of His Word.

How could Moses, born in the ancient Near East, raised in the palace of Pharaoh, educated in the highest courts of Egypt (Acts 7:22), record such knowledge with such unerring clarity? The answer is as humbling as it is profound:

> *"For the prophecy came not in old time by the will of man: but holy men of*

God spake as they were moved by the Holy Ghost."
— 2 Peter 1:21 (KJV)

Moses did not speculate. He did not extrapolate. He did not collect oral traditions and refine them with clever editing. **He received.** Genesis is not the product of philosophical meditation or cultural evolution. It is the result of divine revelation— transmitted by the Spirit of God, through the pen of a prophet, to unveil the origin and order of everything.

Revelation, Not Conjecture

Genesis is not a guess. It is not a poetic approximation or a symbolic allegory, detached from observable reality. It is a revelation—a communication from the infinite mind of God to the finite mind of man. Its truths were not discovered; they were disclosed. Its wisdom was not accumulated; it was bestowed.[44]

---

[44] Craig, William Lane. Reasonable Faith: Christian Truth and Apologetics. Wheaton, IL: Crossway, 2008.

And what is more staggering is that the claims of Genesis are not only theologically sound—they are **scientifically prophetic**. Long before astronomy affirmed the universe had a beginning, Genesis proclaimed, *"In the beginning, God created..."* Long before geology recognized the scars of sudden and global cataclysmic events, Genesis recorded the Flood. Long before paleontology acknowledged the emergence of complex organisms after simpler forms, Genesis set forth that very sequence. Long before linguists traced language divergence back to a singular origin point, Genesis described the confusion of tongues at Babel.

Moses didn't just record the past. He didn't just document what had been. **He revealed what could not have been known**—unless it had been shown.

The Echo of Design

Today, we peer through billion-dollar telescopes, parse genomes with supercomputers, and drill through rock layers that testify of upheaval. And in every one of these discoveries, the design Moses revealed is still speaking.

In every **layer of the earth**, there are whispers of the waters that once covered it.

In every **cell of the body**, there is the imprint of intelligent code—language written in four-letter nucleotides, pulsing with purpose.

In every **language spoken**, there is evidence of a unified origin and a moment of disruption that scattered the sons of men across the earth.

In every **tribe and nation**, there remains a longing for the garden we lost—and a memory, however dim, of the God who once walked with man in the cool of the day.

The message of Genesis is not obsolete. It is not primitive. It is not a relic. It is the **foundation** of truth, both seen and unseen. It proclaims boldly: *This world was not an accident. It was authored.*

Not assumed.
Not imagined.
Revealed—by design.

175

And that design still speaks.

It speaks through the order in nature.
It speaks through the ache in our
souls.
It speaks through the written Word
and the living world alike.
It speaks because **Truth does not
fade—it echoes.**

A Closing Word: From Wonder to Worship

As we close the first book of this journey—*By Design: How Science Proves the Hand of God*—we return to where it all began: "In the beginning God…"

Those four words hold the universe. They demolish randomness. They anchor identity. They ignite purpose. They silence fear.

If there is a beginning, there is a Beginner.

If there is design, there is a Designer.

If there is truth, there is a Truth-Giver.

And if the heavens declare the glory of God (Psalm 19:1), and the Scriptures confirm His authorship, then our response must be more than admiration.

It must be **adoration**.

Consolidated Bibliography

Atkins, Peter. *The Second Law: Energy, Chaos, and Form*. New York: Scientific American Library, 1984.

Austin, Steven A., et al. "Catastrophic Plate Tectonics: A Global Flood Model of Earth History." In *Proceedings of the Third International Conference on Creationism*, 609–621. Pittsburgh: Creation Science Fellowship, 1994.

Bloom, Paul. *Against Empathy: The Case for Rational Compassion*. New York: Ecco, 2016.

Campbell, Lyle. *Historical Linguistics: An Introduction*. Cambridge: MIT Press, 2004.

Cann, Rebecca L., Mark Stoneking, and Allan C. Wilson. "Mitochondrial DNA and Human Evolution." *Nature* 325 (1987): 31–36.

Catling, David C. *Atmospheric Evolution on Inhabited and Lifeless Worlds.* Cambridge: Cambridge University Press, 2017.

Collins, Francis S., et al. "A Vision for the Future of Genomics Research." *Nature* 422 (2003): 835–847.

Craig, William Lane. *Reasonable Faith: Christian Truth and Apologetics.* Wheaton, IL: Crossway, 2008.

Davies, Paul. *The Goldilocks Enigma: Why Is the Universe Just Right for Life?* London: Penguin, 2006.

Efrati, Shai, Amir Hadanny, Yair Fishlev Gabbay, and Yafit Berkovitz. "Hyperbaric oxygen therapy increases telomere length and decreases immunosenescence in isolated blood cells: A prospective trial." *Aging* 13, no. 15 (2020): 20935–20952.

Frazer, James G. *The Golden Bough: A Study in Comparative Religion.* New York: Macmillan, 1922.

Fritz, Paul N. *Flood Legends: Global Clues of a Common Event.* Green Forest, AR: Master Books, 2009.

Gitt, Werner. *In the Beginning Was Information.* Bielefeld, Germany: Christliche Literatur-Verbreitung, 2006.

Guliuzza, Randy J. "Design Analysis Suggests that Human Reproduction is Engineered, Not Evolved." *Acts & Facts* 39, no. 4 (2010): 10–11.

Haidt, Jonathan. *The Righteous Mind: Why Good People Are Divided by Politics and Religion.* New York: Pantheon Books, 2012.

Harari, Yuval Noah. *Sapiens: A Brief History of Humankind.* New York: Harper, 2015.

Hebrews 11:4; John 4:24. These verses are cited directly in the text to support sacrificial and sincere worship.

Heyboer, Marcus, et al. "The Use of Hyperbaric Oxygen Therapy in the Treatment of Radiation-

Induced Soft Tissue Injury: A Review of the Literature." *Undersea & Hyperbaric Medical Society Journal* 44, no. 3 (2017): 191–200.

Jeanson, Nathaniel T. *Replacing Darwin: The New Origin of Species.* Green Forest, AR: Master Books, 2017.

Jeanson, Nathaniel T. *Traced: Human DNA's Big Surprise.* Green Forest, AR: Master Books, 2022.

Kawai, Koichi, et al. "Rib regeneration after costectomy in humans: Histological and radiological studies." *Interdisciplinary Cardiovascular and Thoracic Surgery* 13, no. 3 (2011): 232–236.

LeDoux, Joseph. *Synaptic Self: How Our Brains Become Who We Are.* New York: Viking, 2002.

LeDoux, Joseph. *The Emotional Brain: The Mysterious Underpinnings of Emotional Life.* New York: Simon & Schuster, 1996.

Lovett, Tim. *Noah's Ark: A Feasibility Study*. Green Forest, AR: Master Books, 2008.

Meyer, Stephen C. *Return of the God Hypothesis: Three Scientific Discoveries That Reveal the Mind Behind the Universe*. New York: HarperOne, 2021.

Meyer, Stephen C. *Signature in the Cell: DNA and the Evidence for Intelligent Design*. New York: HarperOne, 2009.

Morris, John D. *The Young Earth: The Real History of the Earth – Past, Present, and Future*. Green Forest, AR: Master Books, 2007.

Nagel, Thomas. *Mind and Cosmos: Why the Materialist Neo-Darwinian Conception of Nature Is Almost Certainly False*. New York: Oxford University Press, 2012.

Paul. Romans 8:21. The concept of the "bondage of corruption" aligns with entropy as a theological and physical principle.

Penzias, Arno A., and Robert W. Wilson. "A Measurement of Excess Antenna Temperature at 4080 Mc/s." *The Astrophysical Journal* 142 (1965): 419–421.

Pinker, Steven. *The Better Angels of Our Nature: Why Violence Has Declined*. New York: Viking, 2011.

Plantinga, Alvin. *Warranted Christian Belief*. New York: Oxford University Press, 2000.

Ross, Hugh. *The Genesis Question: Scientific Advances and the Accuracy of Genesis*. Colorado Springs: NavPress, 2001.

Sagan, Carl. *Cosmos*. New York: Random House, 1980.

Sanford, John C. *Genetic Entropy and the Mystery of the Genome*. Lima, NY: FMS Publications, 2005.

Sapolsky, Robert M. *Behave: The Biology of Humans at Our Best and Worst*. New York: Penguin Press, 2017.

Snelling, Andrew A. *Earth's Catastrophic Past: Geology, Creation & the Flood*. Dallas: Institute for Creation Research, 2009.

Thom, Stephen R., et al. "Stem cell mobilization by hyperbaric oxygen." *American Journal of Physiology-Heart and Circulatory Physiology* 290, no. 4 (2006): H1378–H1386.

Zelickson, Brian D., et al. "Evaluation of the effect of hyperbaric oxygen therapy on collagen synthesis and antioxidant levels in human skin in vivo." *Journal of Investigative Dermatology Symposium Proceedings* 13, no. 1 (2008): 35–39.

Zhang, Pengfei, and Anthony R. Lupo. "The Influence of Water Vapor and Clouds on Earth's Energy Budget." *Atmosphere* 10, no. 4 (2019): 181.

Campbell, Lyle. *Historical Linguistics: An Introduction*. Cambridge: MIT Press, 2004.

Craig, William Lane. *Reasonable Faith: Christian Truth and Apologetics.* Wheaton, IL: Crossway, 2008.

Fritz, Paul N. *Flood Legends: Global Clues of a Common Event.* Green Forest, AR: Master Books, 2009.

Jeanson, Nathaniel T. *Traced: Human DNA's Big Surprise.* Green Forest, AR: Master Books, 2022.

Penzias, Arno A., and Robert W. Wilson. "A Measurement of Excess Antenna Temperature at 4080 Mc/s." *The Astrophysical Journal* 142 (1965): 419–421.

Rees, Martin. *Just Six Numbers: The Deep Forces That Shape the Universe.* New York: Basic Books, 2001.

Sagan, Carl. *Cosmos.* New York: Random House, 1980.

Sanford, John C. *Genetic Entropy and the Mystery of the Genome.* Lima, NY: FMS Publications, 2005.